Computational Electronics

Computational Electronics

Dragica Vasileska and Stephen M. Goodnick

978-3-031-00562-6   paperback   Vasileska/Goodnick

978-3-031-01690-5   ebook   Vasileska /Goodnick

DOI 10.1007/978-3-031-01690-5

A Publication in the Springer series
SYNTHESIS LECTURES ON COMPUTATIONAL ELECTROMAGNETICS
Lecture #6
Editor: Constantine A. Balanis, Arizona State University

Series ISSN Synthesis Lectures on Computational Electromagnetics
Print 1932-1252   Electronic 1932-1716

First Edition
10 9 8 7 6 5 4 3 2 1

Printed in the United States of America

Computational Electronics

Dragica Vasileska and Stephen M. Goodnick
Department of Electrical Engineering, Arizona State University

SYNTHESIS LECTURES ON COMPUTATIONAL ELECTROMAGNETICS #6

ABSTRACT

Computational Electronics is devoted to state of the art numerical techniques and physical models used in the simulation of semiconductor devices from a semi-classical perspective. Computational Electronics, as a part of the general Technology Computer Aided Design (TCAD) field, has become increasingly important as the cost of semiconductor manufacturing has grown exponentially, with a concurrent need to reduce the time from design to manufacture. The motivation for this volume is the need within the modeling and simulation community for a comprehensive text which spans basic drift-diffusion modeling, through energy balance and hydrodynamic models, and finally particle based simulation. One unique feature of this book is a specific focus on numerical examples, particularly the use of commercially available software in the TCAD community. The concept for this book originated from a first year graduate course on Computational Electronics, taught now for several years, in the Electrical Engineering Department at Arizona State University. Numerous exercises and projects were derived from this course and have been included. The prerequisite knowledge is a fundamental understanding of basic semiconductor physics, the physical models for various device technologies such as *pn* diodes, bipolar junction transistors, and field effect transistors.

KEYWORDS

semiconductor device simulation, semiconductor transport, computational science and engineering, integrated circuit technology, technology computer aided design

Contents

C H A P T E R   1

Introduction to Computational Electronics

As semiconductor feature sizes shrink into the nanometer scale regime, even conventional device behavior becomes increasingly complicated as new physical phenomena at short dimensions occur, and limitations in material properties are reached [1]. In addition to the problems related to the understanding of actual operation of ultrasmall devices, the reduced feature sizes require more complicated and time-consuming manufacturing processes. This fact signifies that a pure trial-and-error approach to device optimization will become impossible since it is both too time consuming and too expensive. Since computers are considerably cheaper resources, simulation is becoming an indispensable tool for the device engineer. Besides offering the possibility to test hypothetical devices which have not (or could not) yet been manufactured, simulation offers unique insight into device behavior by allowing the observation of phenomena that cannot be measured on real devices. *Computational Electronics* [2,3] in this context refers to the physical simulation of semiconductor devices in terms of charge transport and the corresponding electrical behavior. It is related to, but usually separate from process simulation, which deals with various physical processes such as material growth, oxidation, impurity diffusion, etching, and metal deposition inherent in device fabrication [4] leading to integrated circuits. Device simulation can be thought of as one component of technology for computer-aided design (TCAD), which provides a basis for device modeling, which deals with compact behavioral models for devices and subcircuits relevant for circuit simulation in commercial packages such as SPICE [5]. The relationship between various simulation design steps that have to be followed to achieve certain customer need is illustrated in Figure 1.1.

The goal of *Computational Electronics* is to provide simulation tools with the necessary level of sophistication to capture the essential physics while at the same time minimizing the computational burden so that results may be obtained within a reasonable time frame. Figure 1.2 illustrates the main components of semiconductor device simulation at any level. There are two main kernels, which must be solved self-consistently with one another, the transport equations governing charge flow, and the fields driving charge flow. Both are coupled strongly to one

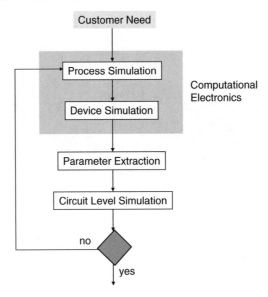

FIGURE 1.1: Design sequence to achieve desired customer need

another, and hence must be solved simultaneously. The fields arise from external sources, as well as the charge and current densities which act as sources for the time varying electric and magnetic fields obtained from the solution of Maxwell's equations. Under appropriate conditions, only the quasi-static electric fields arising from the solution of Poisson's equation are necessary.

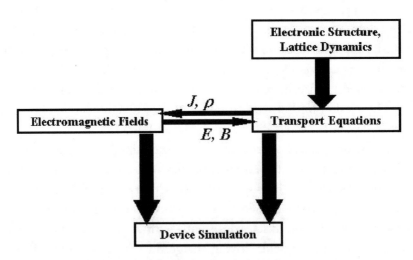

FIGURE 1.2: Schematic description of the device simulation sequence

The fields, in turn, are driving forces for charge transport as illustrated in Figure 1.3 for the various levels of approximation within a hierarchical structure ranging from compact modeling at the top to an exact quantum mechanical description at the bottom. At the very beginning of semiconductor technology, the electrical device characteristics were estimated using simple analytical models (e.g. the gradual channel approximation for MOSFETs) relying on the drift–diffusion (DD) model. Various approximations have to be made to obtain closed-form solutions, but the resulting models captured the basic features of the devices [6]. These approximations include simplified doping profiles and device geometries. With the ongoing refinements and improvements in technology, these approximations lost their basis and a more accurate description was required. This goal could be achieved by solving the DD equations numerically. Numerical simulation of carrier transport in semiconductor devices dates back to the famous work of Scharfetter and Gummel [7], who proposed a robust discretization of the DD equations which is still in use today.

However, as semiconductor devices were scaled into the submicrometer regime, the assumptions underlying the DD model lost their validity. Therefore, the transport models have been continuously refined and extended to more accurately capture transport phenomena occurring in these devices. The need for refinement and extension is primarily caused by the ongoing feature size reduction in state-of-the-art technology. As the supply voltages cannot be scaled accordingly without jeopardizing the circuit performance, the electric field inside the devices increases. A large electric field which rapidly changes over small length scales gives rise to nonlocal and hot-carrier effects which begin to dominate device performance. An accurate description of these phenomena is required and is becoming a primary concern for industrial applications.

To overcome some of the limitations of the DD model, extensions have been proposed which basically add an additional balance equation for the average carrier energy [8]. Furthermore, an additional driving term is added to the current expression which is proportional to the gradient of the carrier temperature. However, a vast number of these models exist, and there is a considerable amount of confusion as to their relation to each other. It is now a common practice in industry to use standard hydrodynamic models in trying to understand the operation of as-fabricated devices, by adjusting any number of phenomenological parameters (e.g., mobility, impact ionization coefficient, etc.). However, such tools do not have predictive capability for ultrasmall structures, for which it is necessary to relax some of the approximations in the Boltzmann transport equation (BTE) [9]. Therefore, one needs to move downward to the quantum transport area in the hierarchical map of transport models shown in Figure 1.3 where, at the very bottom we have the Green's function approach [10–12]. The latter is the most exact, but at the same time the most difficult of all. In contrast to, for example, the Wigner function approach (which is Markovian in time), the Green's functions method allows one to consider

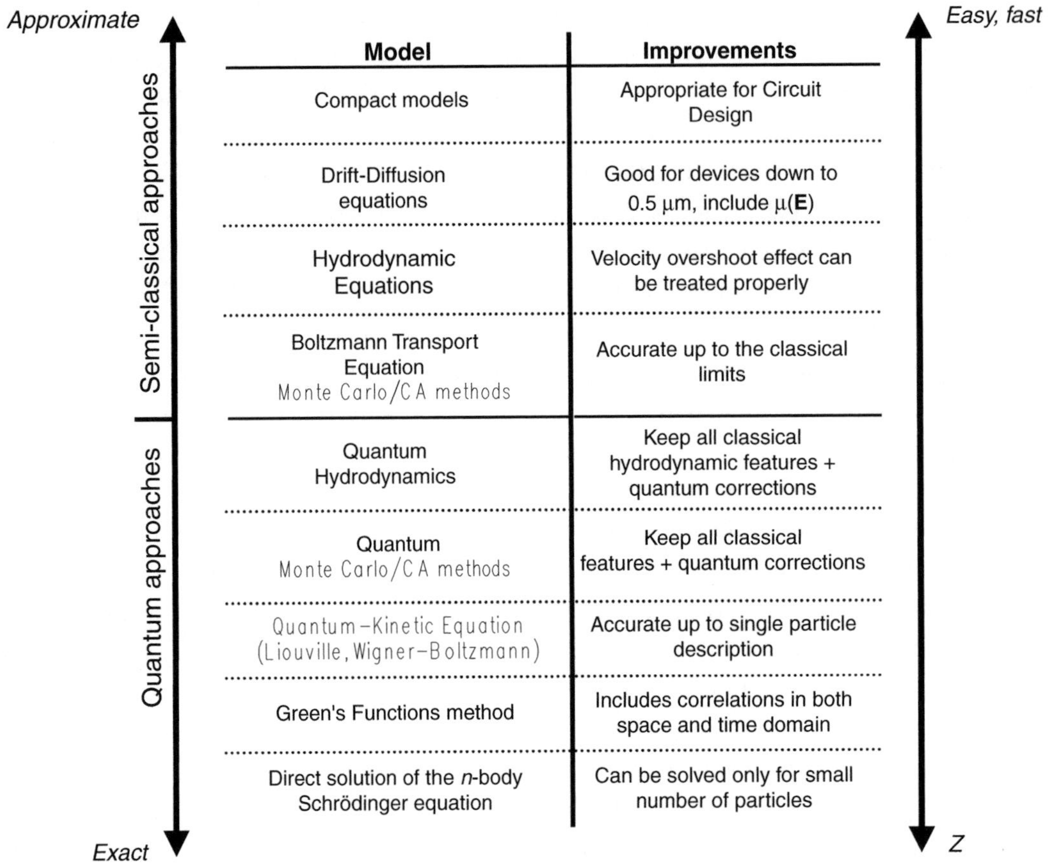

Model	Improvements
Compact models	Appropriate for Circuit Design
Drift-Diffusion equations	Good for devices down to 0.5 μm, include $\mu(\mathbf{E})$
Hydrodynamic Equations	Velocity overshoot effect can be treated properly
Boltzmann Transport Equation Monte Carlo/CA methods	Accurate up to the classical limits
Quantum Hydrodynamics	Keep all classical hydrodynamic features + quantum corrections
Quantum Monte Carlo/CA methods	Keep all classical features + quantum corrections
Quantum–Kinetic Equation (Liouville, Wigner–Boltzmann)	Accurate up to single particle description
Green's Functions method	Includes correlations in both space and time domain
Direct solution of the n-body Schrödinger equation	Can be solved only for small number of particles

Approximate … *Exact* (left axis); *Easy, fast* … *Z* (right axis)

Semi-classical approaches / Quantum approaches

FIGURE 1.3: Illustration of the hierarchy of transport models

simultaneously correlations in space and time, both of which are expected to be important in nanoscale devices. However, the difficulties in understanding the various terms in the resultant equations and the enormous computational burden needed for its actual implementation make the usefulness in understanding quantum effects in actual devices of limited values. For example, the only successful utilization of the Green's function approach commercially is the Nano-Electronics MOdeling (NEMO) simulator [13], which is effectively 1D and is primarily applicable to resonant tunneling diodes.

From the discussion above it follows that, contrary to the recent technological advances, the present state of the art in device simulation is currently lacking in the ability to treat these new challenges in scaling of device dimensions from conventional down to quantum scale devices. For silicon devices with active regions below 0.2 μm in diameter, macroscopic transport descriptions based on DD models are clearly inadequate. As already noted,

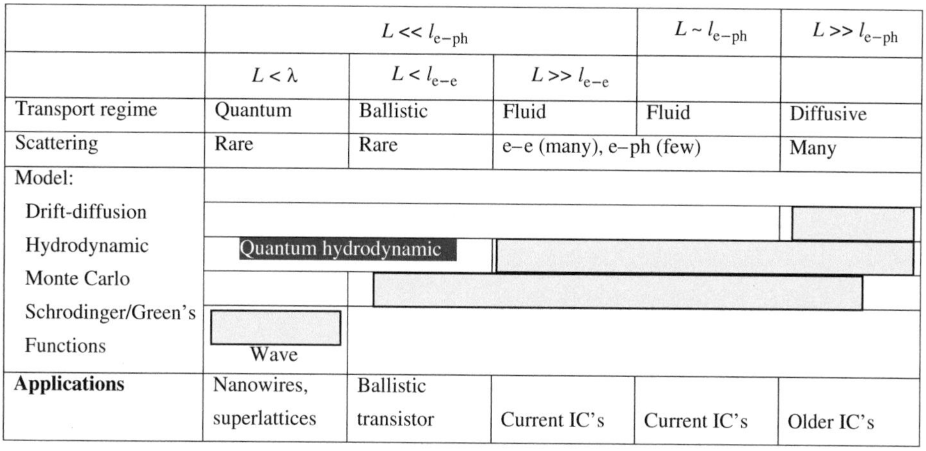

	$L \ll l_{e-ph}$			$L \sim l_{e-ph}$	$L \gg l_{e-ph}$
	$L < \lambda$	$L < l_{e-e}$	$L \gg l_{e-e}$		
Transport regime	Quantum	Ballistic	Fluid	Fluid	Diffusive
Scattering	Rare	Rare	e–e (many), e–ph (few)		Many
Model:					
Drift-diffusion					
Hydrodynamic	Quantum hydrodynamic				
Monte Carlo					
Schrodinger/Green's					
Functions	Wave				
Applications	Nanowires, superlattices	Ballistic transistor	Current IC's	Current IC's	Older IC's

FIGURE 1.4: Relationship between various transport regimes and significant length-scales

even standard hydrodynamic models do not usually provide a sufficiently accurate description since they neglect significant contributions from the tail of the phase space distribution function in the channel regions [14, 15]. Within the requirement of self-consistently solving the coupled transport-field problem in this emerging domain of device physics, there are several computational challenges, which limit this ability. One is the necessity to solve both the transport and the Poisson's equations over the full 3D domain of the device (and beyond if one includes radiation effects). As a result, highly efficient algorithms targeted to high-end computational platforms (most likely in a multiprocessor environment) are required to fully solve even the appropriate field problems. The appropriate level of approximation necessary to capture the proper nonequilibrium transport physics relevant to a future device model is an even more challenging problem both computationally and from a fundamental physics framework.

In this book, we give an overview of the basic techniques used in the field of *Computational Electronics* related to device simulation. The multiple scale transport in doped semiconductors is summarized in Figure 1.4 in terms of the transport regimes, relative importance of the scattering mechanisms and possible applications.

The book is organized as follows. In Chapter 2 we introduce some basic concepts, such as band structure, carrier dynamics, effective masses, etc. In Chapter 3 we introduce the DD model via introduction of the BTE and the relaxation-time approximation. Discretization schemes for the Poisson and the continuity equations are also elaborated in this chapter. The balance equations and their corresponding explicit time discretization schemes are discussed in Chapter 4. Chapter 5 is probably the most useful chapter to the user. In this chapter, although the emphasis is on the usage of the SILVACO simulation software, the discussion presented is very general and the points made are applicable to any device simulation software. Particularly important are the examples given and the hints on how to perform more effective simulations. The Ensemble Monte Carlo (EMC) technique for the solution of the BTE is discussed in Chapter 6, thus completing the main goal of this book to cover the various methods used in semiclassical device simulation. The Appendix A is included to introduce the reader in solving linear algebraic equations of the form $Ax = b$, where A is a sparse matrix and special solution techniques apply. In Appendix B we discuss mobility measurement techniques and the mobility modeling as is usually done in arbitrary device simulator.

CHAPTER 2

Semiconductor Fundamentals

In this Chapter, we provide a brief review of semiconductor physics relevant to the needs of Computational Electronics. We begin with a brief review of the electronic states in a periodic potential as seen by electrons in crystalline semiconductor materials, i.e., the semiconductor bandstructure. We then introduce the important concepts of effective mass and density of states. We then look at transport in semiconductors through the semiclassical Boltzmann Transport Equation (BTE), which is the basis for all the transport simulation methods discussed in the present volume, and look at some simplifying assumptions such as the relaxation-time approximation for its solutions.

2.1 SEMICONDUCTOR BANDSTRUCTURE

The basis for discussing transport in semiconductors is the underlying electronic *band structure* of the material arising from the solution of the many-body Schrödinger equation in the presence of the periodic potential of the lattice, which is discussed in a host of solid state physics textbooks. The solution of the one-particle Schrödinger equation in the presence of the periodic potential of the lattice (and all the other electrons by an effective one-particle potential) are in the form of Bloch functions [16, 17]

$$\psi_{n,\mathbf{k}}(\mathbf{r}) = u_{n,\mathbf{k}}(\mathbf{r})e^{i\mathbf{k}\cdot\mathbf{r}}, \tag{2.1}$$

where $\mathbf{k}$ is the wavevector, and n labels the band index corresponding to different solutions for a given wavevector. The cell-periodic function, $u_{n,\mathbf{k}}(\mathbf{r})$, has the periodicity of the lattice and modulates the traveling wave solution associated with the free particle motion of electrons. The energy eigenvalues, $E_n(\mathbf{k})$, associated with the Bloch eigenfunctions, $\psi_{n,\mathbf{k}}$ above, form what is commonly referred to as the energy bandstructure. The energy, $E_n(\mathbf{k})$, is periodic as a function of $\mathbf{k}$, with a periodicity corresponding to the reciprocal lattice in k-space associated with the real-space lattice. The energy is therefore uniquely specified within the unit cell of this reciprocal lattice, referred to as the first Brillouin zone (BZ1).

In the usual quantum mechanical picture associated with the wave-particle duality of matter, the electron motion through the crystal is visualized as a localized wave-packet composed of a superposition of Bloch states of different wavevectors around an average wavevector, $\mathbf{k}$. The expectation value of the particle velocity then corresponds to the group velocity of this wave packet, or

$$\mathbf{v} = \frac{1}{\hbar} \nabla_\mathbf{k} E_n(\mathbf{k}). \tag{2.2}$$

A brief look at the symmetry properties of the Bloch functions gives some insight into the nature of the bandstructure in semiconductors. First consider the atomic orbitals of the individual atoms that constitute the semiconductor crystal. Typical semiconductors have an average of four valence electrons per atom composed of partially filled s- and p-type orbitals that contribute to bonding, primarily tetrahedral bonds that formed through sp^3 hybridization. The symmetry (or geometric) properties of these atomic orbitals are apparent from consideration of their angular components

$$
\begin{aligned}
s &= 1, \\
p_x &= \frac{x}{r} = \sqrt{3}\sin\theta\cos\varphi, \\
p_y &= \frac{y}{r} = \sqrt{3}\sin\theta\sin\varphi, \\
p_z &= \frac{z}{r} = \sqrt{3}\cos\theta.
\end{aligned}
\tag{2.3}
$$

Let us denote these states by $|S>$, $|X>$, $|Y>$, and $|Z>$. When individual atoms are brought together, these orbitals combine or hybridize into sp^3 molecular orbitals to form covalent bonds composed of lower energy, filled "bonding" molecular orbitals, and unfilled "antibonding" orbitals. The separation in energy between the bonding and antibonding orbital states can be viewed as the fundamental origin of the energy "gap" characteristic of all semiconductors. Once all the atoms coalesce to form a crystal, these molecular orbitals overlap and broaden, leading to the energy bandstructure with gaps and allowed energy bands. The mostly filled valence bands are formed primarily from the bonding orbital states, while the unfilled conduction band is primarily associated with the antibonding states.

For semiconductors, one is typically worried about the bandstructure of the conduction and the valence bands only. It turns out that the states near the band-edges behave very much like the $|S>$ and the three p-type states that they had when they were individual atoms.

Electronic band structure calculation methods can be grouped into two general categories [18]. The first category consists of *ab initio* methods, such as Hartree-Fock or Density Functional Theory (DFT) (see for example Ref. [19]), which calculate the electronic structure from first principles, i.e., without the need for empirical fitting parameters. In general, these methods utilize a variational approach to calculate the ground state energy of a many-body system, where the system is defined at the atomic level. The original calculations were performed on

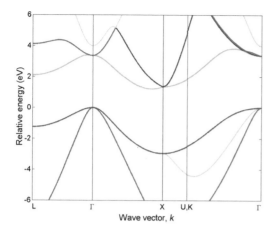
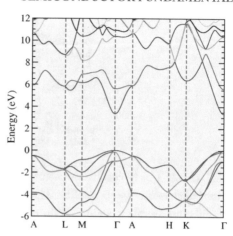

FIGURE 2.1: Empirical pseudopotential calculation of the electronic bandstructure in Si (*left panel*) and wurtzite GaN (*right panel*)

systems containing a few atoms. Today, calculations are performed using thousands of atoms but are computationally expensive, sometimes requiring massively parallel computers.

In contrast to *ab initio* approaches, the second category consists of *empirical* methods, such as the Orthogonalized Plane Wave (OPW) [20], tight-binding [21] (also known as the Linear Combination of Atomic Orbitals (LCAO) method), the $\mathbf{k} \cdot \mathbf{p}$ method [22], and the local [23], or the nonlocal [24] empirical pseudopotential method (EPM). These methods involve empirical parameters to fit experimental data such as the band-to-band transitions at specific high-symmetry points derived from optical absorption experiments. The appeal of these methods is that the electronic structure can be calculated by solving a one-electron Schrödinger wave equation (SWE). Thus, empirical methods are computationally less expensive than *ab initio* calculations and provide a relatively easy means of generating the electronic band structure. Figure 2.1 shows an example of the calculated bandstructure for Si (diamond lattice) and GaN (wurtzite lattice) using the empirical pseudopotential method. In comparing this figure to the simplified bandstructure shown in Figure 2.1 (discussed below), we see that while the basic features are evident such as the indirect bandgap in the Si case and the direct gap in the GaN case, the actual E–$\mathbf{k}$ relationship is quite complicated, with multiple conduction and valence bands and band crossings which make the identification of individual bands somewhat ambiguous.

2.2   SIMPLIFIED BAND STRUCTURE MODELS

In terms of charge transport in semiconductors, it is usually too difficult to deal with the complication of the detailed bandstructure shown in Figure 2.1, and so simplifications are sought. Usually free carriers (electrons or holes) reside at the minimum or maximum of the conduction

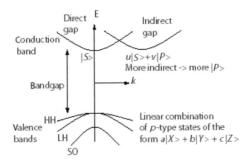

FIGURE 2.2: The typical bandstructure of semiconductors. For direct-gap semiconductors, the conduction band state at $k = 0$ is s-like. The valence band states are linear combinations of p-like orbitals. For indirect-gap semiconductors on the other hand, even the conduction band minima states have some amount of p-like nature mixed into the s-like state

or valence bands, respectively. We see from Figure 2.1 that the E versus $\mathbf{k}$ relation appears quadratic close to an extremum, either concave up or down, which is similar to simply dispersion relation for free electrons quantum mechanically. Depending on the curvature, however, the *effective mass* of the carrier may be smaller or larger than the free electron mass, m_0, and even negative for the case of the valence band, corresponding to holes. Therefore, one often assumes a multiband or multivalley model in which carriers are free electron like, with a unique effective mass for each band or valley. There are usually two levels of approximation used in this case, simple parabolic bands, and nonparabolic bands, in which a correction is included for higher order effects in the dispersion relationship close to an extremum:

(a) Parabolic band

$$E(\mathbf{k}) = \frac{\hbar^2 |\mathbf{k}|^2}{2m_0^*}, \tag{2.4}$$

where m_0^* is the effective mass at the conduction band minimum (or valence band maximum). The particle velocity is simply given from Eq. (2.2) as

$$\mathbf{v} = \frac{1}{\hbar} \nabla_{\mathbf{k}} E(\mathbf{k}) = \frac{\hbar \mathbf{k}}{m_0^*}, \tag{2.5}$$

where the classical momentum and crystal momentum are now identically equal, $\mathbf{p} = \hbar \mathbf{k} = m_0^* \mathbf{v}$.

(b) Nonparabolic band

$$E(\mathbf{k})\left(1 + \alpha E(\mathbf{k})\right) = \frac{\hbar^2 |\mathbf{k}|^2}{2m_0^*}, \tag{2.6}$$

where α is the nonparabolicity factor, which has the dimensions of an inverse energy. The solution of this quadratic equation is

$$E(k) = \frac{\sqrt{1 + \frac{4\alpha\hbar^2|\mathbf{k}|^2}{2m_0^*}} - 1}{2\alpha} \tag{2.7}$$

The velocity is then derived from Eq. (2.2) to be

$$\mathbf{v} = \frac{1}{\hbar}\nabla_k E(\mathbf{k}) = \frac{\hbar\mathbf{k}}{m_0^*}\left(1 + 4\alpha\frac{\hbar^2|\mathbf{k}|^2}{2m_0^*}\right)^{-1/2} = \frac{\hbar\mathbf{k}}{m_0^*[1 + 2\alpha E(\mathbf{k})]}. \tag{2.8}$$

The nonparabolicity factor, α, is related to the degree of admixture of s-like CB states and p-like VB states, given by

$$\alpha = \frac{\left(1 - \frac{m_0^*}{m_0}\right)^2}{E_{\text{gap}}}, \tag{2.9}$$

where $m_0 = 9.11 \times 10^{-31}$ kg is the free electron mass in vacuum, and E_{gap} is the energy gap between valence and conduction band. Hence, smaller bandgap materials have stronger mixing of CB and VB states, and therefore a stronger nonparabolicity.

2.3   CARRIER DYNAMICS

Under the influence of an external field, Bloch electrons in a crystal change their wavevector according to the acceleration theorem

$$\hbar\frac{d\,\mathbf{k}(t)}{dt} = \mathbf{F}, \tag{2.10}$$

where $\mathbf{F}$ is the external force (i.e., external to the crystal field itself) acting on a particle, and $\hbar\mathbf{k}$ plays the role of a pseudo or crystal momentum in the analogy to Newton's equation of motion. The effect on the actual velocity or momentum of the particle is, however, not straightforward as the velocity is related to the group velocity of the wave packet associated with the particle given by Eq. (2.2), where $E_n(\mathbf{k})$ is one of the dispersion relations from a bandstructure calculation such as those shown in Figure 2.1. As the particle moves through k-space under the influence of an electric field, for example, its velocity can be positive or negative, giving the possibility of *Bloch oscillations* if an electron is able to traverse the entire BZ1 without scattering. Only near extremum of the bands, for example at the Γ point in Figure 2.1 for the valence band, or close to the minima in the conduction band, does the dispersion relation resemble that of the free electrons, $E(\mathbf{k}) = \hbar^2 k^2/2m^*$. There, the electron velocity is simply given by $\mathbf{v} = \hbar\mathbf{k}/m^*$, and the momentum is $\mathbf{p} = \hbar\mathbf{k}$, as discussed in the previous section.

In the case of the valence band, the states are nearly full, and current can only be carried by the absence of electrons in a particular state, leading to the concept of *holes*, whose dynamics are identical to that of electrons except their motion is in the opposite direction of electrons, hence they behave as positively charged particles. In relation to transport and device behavior, these holes are then treated as positively charged particles on an equal footing with electrons in the presence of external fields, and in general one has to simulate the motion of both electrons and holes.

For device modeling and simulation, different approximate band models are employed. As long as carriers (electrons and holes) have relatively low energies, they may be treated using the so-called *parabolic band approximation*, where they simply behave as free particles having an effective mass. If more accuracy is desired, corrections due to deviation of the dispersion relation from a quadratic dependence on **k** may be incorporated in the *nonparabolic band model*. If more than one conduction band minimum is important, this model may be extended to a *multivalley model*, where the term valley refers to different conduction minima. Finally, if the entire energy dispersion is used, one usually refers to the model as *full band*.

2.4   EFFECTIVE MASS IN SEMICONDUCTORS

The effective mass of a semiconductor is obtained by fitting the actual $E–k$ diagram around the conduction band minimum or the valence band maximum by a parabola. While this concept is simple enough, the issue turns out to be substantially more complex due to the multitude and the occasional anisotropy of the minima and maxima. In this section we first describe the different relevant band minima and maxima, present the numeric values for germanium, silicon and gallium arsenide and introduce the effective mass for density of states calculations and the effective mass for conductivity calculations.

Most semiconductors can be described as having one band minimum at $k = 0$ as well as several equivalent anisotropic band minima at $k \neq 0$. In addition there are three band maxima of interest which are close to the valence band edge. As an example we consider the simplified band structure of Si shown in Figure 2.3.

The $E–k$ diagram is shown within the first Brillouin zone and along the (1 0 0) direction (see Figure 2.3). The reference of energy is chosen to the edge of the valence band. The lowest band minimum at $k = 0$ directly above the valence band edge occurs at $E_{g,\text{direct}} = 3.2$ eV. This is not the lowest minimum above the valence band edge since there are also six equivalent minima at $k = (x, 0, 0), (-x, 0, 0), (0, x, 0), (0, -x, 0), (0, 0, x),$ and $(0, 0, -x)$ with $x = 5$ nm^{-1}. The minimum energy of all these minima equals 1.12 eV $= E_{g,\text{indirect}}$. The effective mass of these anisotropic minima is characterized by a longitudinal mass along the corresponding equivalent (1 0 0) direction and two transverse masses in the plane perpendicular to the longitudinal direction. In Si, the longitudinal electron mass is $m^*_{e,l} = 0.98\, m_0$ and the transverse electron masses are $m^*_{e,t} = 0.19\, m_0$, where m_0 is the free electron rest mass. Two of the three valence

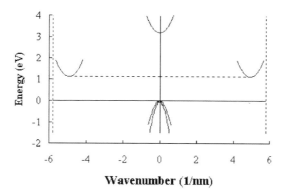

FIGURE 2.3: Simplified E–k diagram for Si within the first Brillouin zone and along the (1 0 0) direction

band maxima occur at 0 eV. These bands are referred to as the light and heavy hole bands with a light hole mass of $m_{\text{lh}}^* = 0.16\ m_0$ and a heavy hole mass of $m_{\text{hh}}^* = 0.46\ m_0$. In addition there is a spin split-off hole band with its maximum at $E_{\text{v,so}} = -0.044$ eV and a split-off hole mass of $m_{\text{v,so}}^* = 0.29\ m_0$ (Figure 2.5).

The values of the energy band minima and maxima as well as the effective masses for germanium, silicon, and gallium arsenide are listed in Table 2.1 below. Figure 2.4 shows the constant energy surfaces in k-space corresponding to the conduction bands of Ge, Si, and GaAs. Figure 2.5 shows the constant energy surfaces for heavy holes, light holes and the spin split off band for Si, illustrating the warped nature of the bands.

The effective mass for density of states calculations (see Table 2.2 below) equals the mass which provides the density of states using the expression for one isotropic maximum or minimum or

$$g_C(E) = \frac{8\pi\sqrt{2}}{h^3} m_e^{*3/2}\sqrt{E - E_C}, \quad \text{for} \quad E \geq E_C \tag{2.11}$$

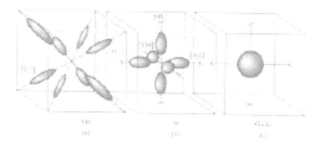

FIGURE 2.4: Constant energy surfaces of the conduction band of Ge, Si, and GaAs. Note that in the case of Ge we have four conduction band minima (since the band minima occurs on the edge of the BZ1), in the case of Si we have six conduction band equivalent valleys and in the case of GaAs we have only one constant energy surface at the center of the Brillouin zone

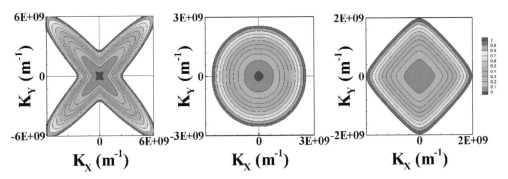

FIGURE 2.5: 3D equi-energy surfaces of heavy hole, light hole and split off band in Si for $k_z = 0$

TABLE 2.1: Values of the Energy Band Minima and Maxima and the Effective Masses for Germanium, Silicon, and Gallium Arsenide

NAME	SYMBOL	GERMANIUM	SILICON	GALLIUM ARSENIDE
Band minimum at $k = 0$				
Minimum energy	$E_{g,\text{direct}}$ [eV]	0.8	3.2	1.424
Effective mass	m_e^*/m_0	0.041	?0.2?	0.067
Band minimum *not* at $k = 0$				
Minimum energy	$E_{g,\text{indirect}}$ [eV]	0.66	1.12	1.734
Longitudinal effective mass	$m_{e,l}^*/m_0$	1.64	0.98	1.98
Transverse effective mass	$m_{e,t}^*/m_0$	0.082	0.19	0.37
Wavenumber at minimum	k [1/nm]	–	–	–
Longitudinal direction		(111)	(100)	(111)
Heavy hole valence band maximum at $E = k = 0$				
Effective mass	m_{hh}^*/m_0	0.28	0.49	0.45
Light hole valence band maximum at $k = 0$				
Effective mass	m_{lh}^*/m_0	0.044	0.16	0.082
Split-off hole valence band maximum at $k = 0$				
Split-off band valence band energy	$E_{v,\text{so}}$ [eV]	−0.028	−0.044	−0.34
Effective mass	$m_{h,\text{so}}^*/m_0$	0.084	0.29	0.154
$m_0 = 9.11 \times 10^{-31}$ kg is the free electron rest mass.				

for the density of states in the conduction band, and

$$g_V(E) = \frac{8\pi\sqrt{2}}{h^3} m_h^{*3/2} \sqrt{E_V - E}, \quad \text{for} \quad E \leq E_V \quad (2.12)$$

for the density of states in the valence band. For instance, for a single band minimum described by a longitudinal mass and two transverse masses, the effective mass for density of states calculations is the geometric mean of the three masses. Including the fact that there are several equivalent minima at the same energy one obtains the effective mass for density of states calculations from

$$m_{e,dos}^* = M_C^{2/3}(m_l m_t m_t)^{1/3}, \quad (2.13)$$

where M_C is the number of equivalent band minima. For silicon one obtains

$$m_{e,dos}^* = (m_l m_t m_t)^{1/3} = (6)^{2/3}(0.89 \times 0.19 \times 0.19)^{1/3} m_0 = 1.08\, m_0. \quad (2.14)$$

The effective mass for conductivity calculations (see Table 2.2 below) is the mass which is used in conduction-related problems accounting for the detailed structure of the semiconductor. These calculations include parameters such as the mobility and diffusion constants. Another example is the calculation of the shallow impurity levels using a hydrogen-like model. As the conductivity of a material is inversely proportional to the effective masses, one finds that the conductivity due to multiple band maxima or minima is proportional to the sum of the inverse of the individual masses, multiplied by the density of carriers in each band, as each maximum or minimum adds to the overall conductivity. For anisotropic minima containing one longitudinal and two transverse effective masses, one has to sum over the effective masses in the different minima along the equivalent directions. The resulting effective mass for bands which have ellipsoidal constant energy surfaces is given by

$$m_{e,cond}^* = \frac{3}{\frac{1}{m_l} + \frac{1}{m_t} + \frac{1}{m_t}} \quad (2.15)$$

provided the material has an isotropic conductivity as is the case for cubic materials. For instance electrons in the X minima of silicon have an effective conductivity mass given by

$$\begin{aligned} m_{e,cond}^* &= 3 \times (1/m_l + 1/m_t + 1/m_t)^{-1} \\ &= 3 \times (1/0.89 + 1/0.19 + 1/0.19)^{-1} m_0 = 0.26\, m_0. \end{aligned} \quad (2.16)$$

2.5 SEMICLASSICAL TRANSPORT THEORY

To completely model the behavior of a semiconductor device, one must know the state of each carrier within the device and their motion, which is the role of transport theory. If carriers are considered as classical particles, one-way of modeling carrier dynamics is to solve Newton's

TABLE 2.2: Effective Mass and Energy Bandgap of Ge, Si, and GaAs

NAME	SYMBOL	GERMANIUM	SILICON	GALLIUM ARSENIDE
Smallest energy bandgap at 300 K	E_g (eV)	0.66	1.12	1.424
Effective mass for density of states calculations				
Electrons	$m^*_{e,dos}/m_0$	0.56	1.08	0.067
Holes	$m^*_{h,dos}/m_0$	0.29	0.57/0.81[1]	0.47
Effective mass for conductivity calculations				
Electrons	$m^*_{e,cond}/m_0$	0.12	0.26	0.067
Holes	$m^*_{h,cond}/m_0$	0.21	0.36/0.386 [25]	0.34

$m_0 = 9.11 \times 10^{-31}$ kg is the free electron rest mass.

equations

$$\frac{d\mathbf{p}}{dt} = -e\mathbf{E} + R(\mathbf{r}, \mathbf{p}, t) \quad \text{and} \quad \mathbf{v}(t) = \frac{d\mathbf{r}}{dt}, \tag{2.17}$$

where $R(\mathbf{r}, \mathbf{p}, t)$ is a random force function due to various random scattering processes such as impurities, lattice vibrations, other particles, etc. which randomly change the energy and momentum of a particle. This approach is the basis of the so-called Langevin equation which most famously was used to describe Brownian motion.

An alternative approach (the so-called kinetic equation approach) is to calculate statistically the probability of finding a carrier with crystal momentum $\mathbf{k}$ at position $\mathbf{r}$ at time t, represented by the distribution function $f(\mathbf{r}, \mathbf{k}, t)$. This approach is the basis of the BTE [26–28], discussed in Section 2.6. Therefore, once the distribution function is specified, various moments of the distribution function can give us particle density, current density, energy density, etc. More precisely

$$n(\mathbf{r}, t) = \frac{1}{V} \sum_k f(\mathbf{r}, \mathbf{k}, t), \text{ particle density,} \tag{2.18}$$

$$\mathbf{J}(\mathbf{r}, t) = -\frac{e}{V} \sum_k \mathbf{v}(\mathbf{k}) f(\mathbf{r}, \mathbf{k}, t), \text{ current density,} \tag{2.19}$$

$$W(\mathbf{r}, t) = \frac{1}{V} \sum_k E(\mathbf{k}) f(\mathbf{r}, \mathbf{k}, t), \text{ energy density.} \tag{2.20}$$

A full quantum-mechanical view to this problem is rather difficult [29,30]. The uncertainty principle states, for example, that we cannot specify simultaneously the position and the momentum of the particle. Hence, one needs to adopt a coarse-grained average point of view, in which positions are specified within a macroscopic volume, and momenta are also specified within some interval. If one tries to go straightforwardly and construct $f(\mathbf{r}, \mathbf{k}, t)$ from the quantum-mechanical wavefunctions, difficulties arise since f is not necessarily positive definite.

2.5.1   Approximations Made for the Distribution Function

One of the main problems in device analysis is calculation of the distribution function, $f(\mathbf{r}, \mathbf{k}, t)$. Before discussing the formal derivation of $f(\mathbf{r}, \mathbf{k}, t)$ semiclassically from the BTE in the next section, we consider here approximations to the distribution function that are often employed. The two most commonly used approaches are

- Quasi-Fermi level concept.
- Displaced Maxwellian approximation for the distribution function.

(A) Quasi-Fermi level concept

Under equilibrium conditions $np = n_i^2$, where n is the electron concentration, p is the hole concentration and n_i is the intrinsic carrier concentration which follows from the use of the equilibrium Fermi-Dirac distribution functions for electrons and holes (with E_F the Fermi level), i.e.,

$$f_n(E) = \frac{1}{1 + \exp\left(\frac{E - E_F}{k_B T}\right)}, \quad f_p(E) = 1 - f_n(E) = \frac{1}{1 + \exp\left(\frac{E_F - E}{k_B T}\right)}. \tag{2.21}$$

Under nonequilibrium conditions, it may still be useful to represent the distribution functions for electrons and holes by introducing the quasi-Fermi levels, E_{Fn} and E_{Fp}, as

$$f_n(E) = \frac{1}{1 + \exp\left(\frac{E - E_{Fn}}{k_B T}\right)} \quad \text{and} \quad f_p(E) = 1 - f_n(E) = \frac{1}{1 + \exp\left(\frac{E_{Fp} - E}{k_B T}\right)}. \tag{2.22}$$

Therefore, under nonequilibrium conditions and assuming nondegenerate statistics, we have

$$n = N_C \exp\left(\frac{E_{Fn} - E_C}{k_B T}\right) \quad \text{and} \quad p = N_V \exp\left(\frac{E_V - E_{Fp}}{k_B T}\right), \tag{2.23}$$

where N_C and N_V are the effective density of states of the conduction and valence band, respectively [31,32]. The product

$$np = n_i^2 \exp\left(\frac{E_{Fn} - E_{Fp}}{k_B T}\right), \tag{2.24}$$

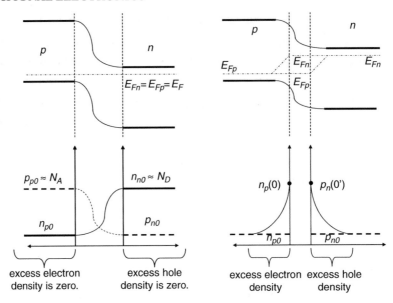

FIGURE 2.6: Energy band profile of a *pn*-diode under equilibrium and nonequilibrium conditions. Note that in order to obtain excess electron density (*bottom right panel*), the electron quasi-Fermi level must move up (*top right panel*), thus increasing the probability of state occupancy. The same is true for the excess hole concentration, where the hole quasi-Fermi level moves downward

suggests that the difference $E_{\mathrm{F}n} - E_{\mathrm{F}p}$ is a measure for the deviation from the equilibrium of the carrier concentrations spatially. However, this cannot be the correct distribution function in k-space since it is even in k, which implies no current flow in the device based on Eq. (2.19) (i.e., the integral over all space of an even function times an odd function is zero by symmetry). However, if the average carrier velocities are much smaller than the thermal velocity (i.e., the spread in velocity), given by $\sqrt{2k_{\mathrm{B}}T/m^*} \approx 10^7$ cm s^{-1} for $m^* = m_0$ (free electron mass), then the approximation of the distribution function by a parameterized Fermi-Dirac function is not so unreasonable (Figure 2.6).

(B) Displaced Maxwellian approximation
A better approximation for the distribution function $f(\mathbf{r}, \mathbf{k}, t)$ is to assume that the distribution function retains its symmetric shape, but that its average momentum is displaced from the origin to account for the net velocity in the direction of the electric field. For example, a particularly suitable form to use is the so-called displaced Maxwellian distribution [33] (Figure 2.7)

$$f(\mathbf{r}, \mathbf{k}, t) = \exp\left(\frac{E_{\mathrm{F}n} - E_{\mathrm{C}0}}{k_{\mathrm{B}}T}\right) \exp\left(-\frac{\hbar^2}{2m^*k_{\mathrm{B}}T}|\mathbf{k} - \mathbf{k}_{\mathrm{d}}|^2\right). \qquad (2.25)$$

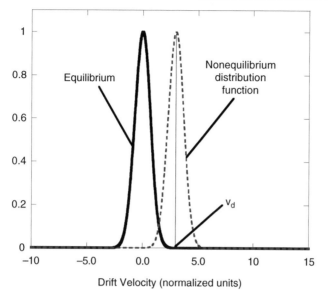

FIGURE 2.7: Displaced Maxwellian distribution function

Using this form of the distribution function gives

$$n(\mathbf{r}, t) = \frac{1}{V} \sum_k f(\mathbf{r}, \mathbf{k}, t) = N_C \exp\left(\frac{E_{Fn} - E_{C0}}{k_B T}\right). \tag{2.26}$$

In the same manner, one finds that the kinetic energy density per carrier is given by

$$u(\mathbf{r}, t) = \frac{1}{2} m^* v_d^2 + \frac{3}{2} k_B T. \tag{2.27}$$

The first term on the RHS represents the drift energy due to average drift velocity, and the second term is the well-known thermal energy term mentioned above associated primarily with the interaction of charge carriers with the lattice through scattering [34].

Since in both cases, the assumptions for the nonequilibrium distribution has been guided by the form of the equilibrium distribution, they are only valid in near-equilibrium conditions. For far-from-equilibrium conditions, the shape of the distribution function can be rather different [35]. This necessitates the solution of the BTE, discussed in the next section.

2.6   BOLTZMANN TRANSPORT EQUATION (BTE)

To derive the BTE consider a region of phase space about the point (x, y, z, p_x, p_y, p_z). The number of particles entering this region in time dt is equal to the number which were in the region of phase space $(x - v_x dt, y - v_y dt, z - v_z dt, p_x - F_x dt, p_y - F_y dt, p_z - F_z dt)$ at a time dt earlier. If $f(x, y, z, p_x, p_y, p_z)$ is the distribution function which relates to the number of particles per unit volume of phase space around this point, then the change df which occurs during time dt due to the motion of the particles in coordinate space and due to the fact that force fields acting on the particles tend to move them from one region to another in momentum space is [36]:

$$df = f(x - v_x dt, y - v_y dt, z - v_z dt, p_x - F_x dt, p_y - F_y dt, p_z - F_z dt)$$
$$- f(x, y, z, p_x, p_y, p_z). \tag{2.28}$$

Using Taylor series expansion, we get

$$\frac{df}{dt} = -\mathbf{v} \cdot \nabla_{\mathbf{r}} f - \mathbf{F} \cdot \nabla_{\mathbf{p}} f. \tag{2.29}$$

So far, only the change in the distribution function due to the motion of particles in coordinate space and due to the momentum changes arising from the force fields acting on the particles have been accounted for. Particles may also be transferred into or out of a given region in phase space by collisions or scattering interactions involving other particles of the distribution or scattering centers external to the assembly of particles under consideration. If the rate of change of the distribution function due to collisions, or scattering, is denoted by $(\partial f / \partial t)_{\text{coll}}$, the total rate of change of f becomes

$$\frac{df}{dt} = -\mathbf{v} \cdot \nabla_{\mathbf{r}} f - \mathbf{F} \cdot \nabla_{\mathbf{p}} f + \left. \frac{\partial f}{\partial t} \right|_{\text{coll}} + s(r, p, t). \tag{2.30}$$

The last term on the RHS of Eq. (2.30), $s(r, p, t)$, represents the change due to generation–recombination processes.

Typically in the semiclassical picture of transport, we deal with a phase space in terms of $\mathbf{r}$ and $\mathbf{k}$, associated with the crystal momentum, rather than the actual particle momentum, $\mathbf{p}$. If we therefore convert from $\mathbf{p}$ to $\mathbf{k}$, replace $\mathbf{v}$ with its expectation value, Eq. (2.2), and take the first two terms on the RHS to the left, Eq. (2.30) becomes

$$\frac{\partial f(\mathbf{r}, \mathbf{k}, t)}{\partial t} + \frac{1}{\hbar} \nabla_{\mathbf{k}} E(\mathbf{k}) \cdot \nabla_{\mathbf{r}} f(\mathbf{r}, \mathbf{k}, t) + \frac{\mathbf{F}}{\hbar} \cdot \nabla_{\mathbf{k}} f(\mathbf{r}, \mathbf{k}, t) = \left. \frac{\partial f}{\partial t} \right|_{\text{coll}} + s(r, k, t). \tag{2.31}$$

Equation (2.31) is the BTE, which is nothing more than a detailed balance of particle flow, or continuity equation, in 6D phase space, as illustrated in Figure 2.8.

The various terms that appear in Eq. (2.31) represent

- $(\partial f / \partial t)_{\text{forces}} = \frac{\mathbf{F}}{\hbar} \cdot \nabla_{\mathbf{k}} f(\mathbf{r}, \mathbf{k}, t)$, where $\mathbf{F} = -\frac{d\mathbf{p}}{dt} = \hbar \frac{d\mathbf{k}}{dt} = q(\mathbf{E} + \mathbf{v} \times \mathbf{B})$, the total force equals the sum of the force due to the electric field and the Lorentz force due to the magnetic flux density, $\mathbf{B}$.

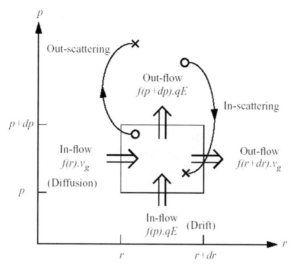

FIGURE 2.8: A cell in two-dimensional phase space. The three processes, namely drift, diffusion, and scattering, that affect the evolution of $f(r, p, t)$ with time in phase space are shown

- $(\partial f/\partial t)_{\text{diff}} = -\frac{1}{\hbar}\nabla_{\mathbf{k}}E(\mathbf{k}) \cdot \nabla_{\mathbf{r}}f$. This term arises if there is a spatial variation in the distribution function due to concentration or temperature gradients, both of which will result in a diffusion of carriers in coordinate space.

- $(\partial f/\partial t)_{\text{coll}}$ is the collision term which equals the difference between the in-scattering and the out-scattering processes, given by the collision integral (sub), i.e.,

$$\left(\frac{\partial f}{\partial t}\right)_{\text{coll}} = \sum_{k'}\{S(\mathbf{k}', \mathbf{k})f(\mathbf{k}')[1 - f(\mathbf{k})] - S(\mathbf{k}, \mathbf{k}')f(\mathbf{k})[1 - f(\mathbf{k}')]\} = \hat{C}f,$$

(2.32)

where the spatial coordinate is understood. The presence of $f(\mathbf{k})$ and $f(\mathbf{k}')$ in the collision integral makes the BTE rather complicated integro-differential equation for $f(\mathbf{r}, \mathbf{k}, t)$, whose solution requires a number of simplifying assumptions. In the absence of perturbing fields and temperature gradients, the distribution function must be the equilibrium Fermi-Dirac function. In this case, the collision term must vanish and the principle of detailed balance gives for all $\mathbf{k}$ and $\mathbf{k}'$ and all scattering mechanisms

$$\frac{S(\mathbf{k}, \mathbf{k}')}{S(\mathbf{k}', \mathbf{k})} = \frac{f_0(\mathbf{k}')[1 - f_0(\mathbf{k})]}{f_0(\mathbf{k})[1 - f_0(\mathbf{k}')]}.$$

(2.33)

Therefore, if the phonons interacting with the electrons are in thermal equilibrium, we get

$$\frac{S(\mathbf{k}, \mathbf{k}')}{S(\mathbf{k}', \mathbf{k})} = \exp\left(\frac{E_{\mathbf{k}} - E_{\mathbf{k}'}}{k_B T}\right).$$

(2.34)

This relation must be satisfied regardless of the origin of the scattering forces. If, for example, we assume $E_\mathbf{k} > E_{\mathbf{k}'}$, then $S(\mathbf{k}, \mathbf{k}')$ which involves emission must exceed $S(\mathbf{k}, \mathbf{k}')$ which involves absorption. Note that the BTE is valid under assumptions of semiclassical transport. Some of these include that the energy band picture holds even under very high fields, that collisions are instantaneous and memoryless (i.e., no dependence on initial conditions), and that the one-electron picture holds, i.e., that higher orders or correlation in the electron–electron motion are neglected. The phonons are usually treated as in equilibrium as well, although the condition of nonequilibrium phonons may be included through an additional kinetic equation for the phonons as well [37].

2.7 SCATTERING PROCESSES

Free carriers (electrons and holes) interact with the crystal and with each other through a variety of scattering processes which relax the energy and momentum of the particle. Based on first order, time-dependent perturbation theory, the transition rate from an initial state $\mathbf{k}$ in band n, to a final state $\mathbf{k}'$ in band m for the jth scattering mechanism is given by Fermi's Golden rule [38]

$$\Gamma_j[n, \mathbf{k}; m, \mathbf{k}'] = \frac{2\pi}{\hbar} |\langle m, \mathbf{k}'|V_j(\mathbf{r})|n, \mathbf{k}\rangle|^2 \delta(E_{\mathbf{k}'} - E_\mathbf{k} \mp \hbar\omega), \qquad (2.35)$$

where $V_j(\mathbf{r})$ is the scattering potential of this process, $E_\mathbf{k}$ and $E_{\mathbf{k}'}$ are the initial and final state energies of the particle. The delta function describes conservation of energy, valid for long times after the collision is over, with $\hbar\omega$ the energy absorbed (upper sign) or emitted (lower sign) during the process. The total rate of scattering for a particle in an initial state $\mathbf{k}$ in band n to any possible final state is given by an integral (sum) over all final wavevectors (bands)

$$\Gamma_j[n, \mathbf{k}] = \frac{2\pi}{\hbar} \sum_{m, \mathbf{k}'} |\langle m, \mathbf{k}'|V_j(\mathbf{r})|n, \mathbf{k}\rangle|^2 \delta(E_{\mathbf{k}'} - E_\mathbf{k} \mp \hbar\omega). \qquad (2.36)$$

There are major limitations to the use of the Golden rule due to effects such as *collision broadening* and *finite collision duration time* [86]. The energy conserving delta function is only valid asymptotically for times long after the collision is complete. The broadening in the final state energy is given roughly by $\Delta E \approx \hbar/\tau$, where τ is the time after the collision, which implies that the normal $E(\mathbf{k})$ relation is only recovered at long times. Attempts to account for such *collision broadening* in Monte Carlo simulation have been reported in the literature [39, 40], although this is still an open subject of debate. Inclusion of the effects of *finite collision duration* in Monte Carlo simulation have also been proposed [41, 42]. Beyond this, there is still the problem of dealing with the quantum mechanical phase coherence of carriers, which is neglected in the scatter free-flight algorithm of the Monte Carlo algorithm, and goes beyond the semiclassical BTE description.

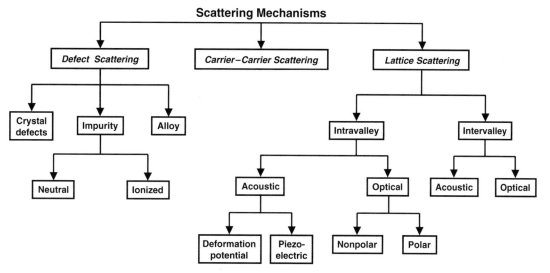

FIGURE 2.9: Scattering mechanisms in a typical semiconductor

Figure 2.9 lists the typical scattering mechanisms present in a semiconductor. They are roughly divided into scattering due to crystal defects, which is primarily elastic in nature, lattice scattering between electrons (holes) and lattice vibrations or phonons, which is inelastic, and finally scattering between the particles themselves, including both single particle and collective type excitations. Phonon scattering involves different modes of vibration, either acoustic or optical, as well as both transverse and longitudinal modes. Carriers may either emit or absorb quanta of energy from the lattice, in the form of phonons, in individual scattering events. The designation of inter- versus intra-valley scattering comes from the multivalley band-structure model of semiconductors discussed in Section 2.2, and refers to whether the initial and final states are in the same valley or in different valleys.

2.8   RELAXATION-TIME APPROXIMATION

Analytical solutions of the Boltzmann equation are possible only under very restrictive assumptions [43]. Direct numerical methods for device simulation have been limited by the complexity of the equation, which in the complete 3D time-dependent form requires seven independent variables for time, space, and momentum. In recent times, more powerful computational platforms have spurred a renewed interest in numerical solutions based on the spheroidal harmonics expansion of the distribution function [44]. To date, most direct solutions of the BTE in semiconductor applications have been based on stochastic solution methods (Monte Carlo), which involve the simulation of particle trajectories rather than the direct solution of partial differential equations, as discussed in detail in Chapter 6.

Most conventional device simulations are based on approximate models for transport which are derived from the Boltzmann equation, and coupled to Poisson's equation for self-consistency. In the simplest approach, the relaxation-time approximation is invoked, where the total distribution function is split into a symmetric part in terms of the momentum (which is generally large) and an asymmetric term in the momentum (which is small). In other words,

$$f(\mathbf{r}, \mathbf{k}, t) = f_S(\mathbf{r}, \mathbf{k}, t) + f_A(\mathbf{r}, \mathbf{k}, t). \tag{2.37}$$

Then, for nondegenerate semiconductors $(1 - f) \approx 1$, the collision integral Eq. (2.32) may be written as

$$\left(\frac{\partial f}{\partial t}\right)_{\text{coll}} = \sum_{\mathbf{k}'} [f(\mathbf{k}')S(\mathbf{k}', \mathbf{k}) - f(\mathbf{k})S(\mathbf{k}, \mathbf{k}')]$$

$$= \underbrace{\sum_{\mathbf{k}'} [f_S(\mathbf{k}')S(\mathbf{k}', \mathbf{k}) - f_S(\mathbf{k})S(\mathbf{k}, \mathbf{k}')]}_{(\partial f_S/\partial t)_{\text{coll}}} + \underbrace{\sum_{\mathbf{k}'} [f_A(\mathbf{k}')S(\mathbf{k}', \mathbf{k}) - f_A(\mathbf{k})S(\mathbf{k}, \mathbf{k}')]}_{(\partial f_A/\partial t)_{\text{coll}}}.$$

$$\tag{2.38}$$

We now consider two cases:

a) Equilibrium conditions: $f_S = f_0$, $f_A = 0 \rightarrow (\frac{\partial f}{\partial t})_{\text{coll}} = (\frac{\partial f_S}{\partial t})_{\text{coll}} = 0$.

b) Nonequilibrium conditions when $f_A \neq 0$. In this case, we must consider two different situations:
  - Low-field conditions, where f_S retains its equilibrium form with $T_C = T_L$. In this case $(\partial f_S/\partial t)_{\text{coll}} = 0$.
  - High-field conditions when $T_C \neq T_L$ and f_S does not retain its equilibrium form. In this case $(\partial f_S/\partial t)_{\text{coll}} \neq 0$.

In all of these cases, a plausible form for the term $(\partial f_A/\partial t)_{\text{coll}}$ is

$$\left(\frac{\partial f_A}{\partial t}\right)_{\text{coll}} = -\frac{f_A}{\tau_f}, \tag{2.39}$$

where τ_f is a characteristic time that describes how the distribution function relaxes to its equilibrium form. With the above discussion, we may conclude that

  - At low fields: $(\frac{\partial f}{\partial t})_{\text{coll}} = (\frac{\partial f_A}{\partial t})_{\text{coll}} = -\frac{f_A}{\tau_f}$,
  - At high fields: $(\frac{\partial f}{\partial t})_{\text{coll}} = (\frac{\partial f_S}{\partial t})_{\text{coll}} + (\frac{\partial f_A}{\partial t})_{\text{coll}} = (\frac{\partial f_S}{\partial t})_{\text{coll}} - \frac{f_A}{\tau_f}$.

To understand the meaning of the relaxation time, we consider a semiconductor in which there are no spatial and momentum gradients. With the gradient terms zero, the BTE becomes

$$\frac{\partial f}{\partial t} = \left(\frac{\partial f_A}{\partial t}\right)_{scatt} = -\frac{f_A}{\tau_f} = -\frac{f - f_0}{\tau_f}, \tag{2.40}$$

i.e.,

$$\frac{\partial f}{\partial t} + \frac{f}{\tau_f} = \frac{f_0}{\tau_f}. \tag{2.41}$$

The solution of this first-order differential equation is

$$f(t) = f_0 + [f(0) - f_0]e^{-t/\tau_f}. \tag{2.42}$$

This result suggests that any perturbation in the system will decay exponentially with a characteristic time constant τ_f. It also suggests that the RTA is only good when $[f(0) - f_0]$ is not very large. Note that an important restriction for the relaxation-time approximation to be valid is that τ_f is independent of the distribution function and the applied electric field.

2.9   SOLVING THE BTE IN THE RELAXATION-TIME APPROXIMATION

Let us consider the simple case of a uniformly doped semiconductor with a constant electric field throughout. Since there are no spatial gradients, $\nabla_r f = 0$. Under steady-state conditions we also have $\partial f / \partial t = 0$. With the above simplifications, the BTE reduces to

$$\mathbf{F} \cdot \nabla_{\mathbf{p}} f = \frac{1}{\hbar} \mathbf{F} \cdot \nabla_{\mathbf{k}} f = \left(\frac{\partial f}{\partial t}\right)_{coll}. \tag{2.43}$$

For parabolic bands and choosing the coordinate system such that the electric field is along the z-axis, one can expand the distribution function into Legendre polynomials

$$f(z, p) = f_0(z, p) + \sum_{n=1}^{\infty} g_n(E) P_n(\cos \theta), \tag{2.44}$$

where $P_0 = 1$, $P_1 = \cos \theta$, $P_2 = \frac{3}{2} \cos^2 \theta - \frac{1}{2}, \ldots$. In the above expressions, θ is the angle between the applied field (along the symmetry axis), and the momentum of the carriers. For sufficiently low fields, we expect that only the lowest order term is important, so that

$$f(p) \cong f_0(p) + g_1(p) \cos \theta = f_0(p) + f_A(p). \tag{2.45}$$

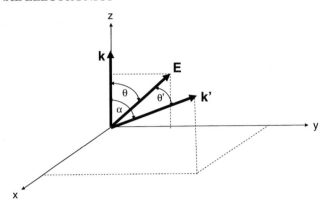

FIGURE 2.10: Coordinate system for scattering relative to an applied electric field

Substituting the results on the LHS of the BTE and using parabolic dispersion relation gives

$$\text{LHS} = -\frac{e}{\hbar}\mathbf{E} \cdot \nabla_\mathbf{k}[f_0(p) + g_1(p)\cos\theta]$$
$$\approx -\frac{e}{\hbar}\mathbf{E} \cdot \nabla_\mathbf{k} f_0(p) = -e\mathbf{E} \cdot \mathbf{v}\frac{\partial f_0}{\partial\varepsilon} = -eEv\cos\theta\frac{\partial f_0}{\partial\varepsilon},$$

(2.46)

where, as previously noted θ is the angle between the electric field and $\mathbf{v}$.

We now consider the collision integral on the RHS of the BTE given in Eq. (2.43). Substituting the first order approximation for $f(\mathbf{p})$ gives

$$\frac{\partial f}{\partial t}\Big|_{\text{coll}} = \sum_{k'}\big[S(k', k)f_0(k') - S(k, k')f_0(k)\big]$$
$$+ \sum_{k'}\big[S(k', k)g_1(k')\cos\theta' - S(k, k')g_1(k)\cos\theta\big]$$
$$= -g_1(k)\cos\theta\sum_{k'}S(k, k')\left[1 - \frac{S(k', k)g_1(k')\cos\theta'}{S(k, k')g_1(k)\cos\theta}\right]$$
$$= -g_1(k)\cos\theta\sum_{k'}S(k, k')\left[1 - \frac{f_0(k)g_1(k')\cos\theta'}{f_0(k')g_1(k)\cos\theta}\right],$$

(2.47)

where in the last line of the derivation we have used the principle of detailed balance. We can further simplify the above result, by considering the following coordinate system (Figure 2.10).

Within this coordinate system, we have

$$k = (0, 0, k),$$
$$k' = (k'\sin\alpha\cos\varphi, k'\sin\alpha\sin\varphi, k'\cos\alpha),$$
$$E = (0, E\sin\theta, E\cos\theta),$$

(2.48)

which leads to

$$E \cdot k' = Ek' \cos \theta' = Ek' (\sin \alpha \sin \varphi \sin \theta + \cos \alpha \cos \theta). \tag{2.49}$$

The integration over φ will make this term vanish, and under these circumstances we can write

$$\frac{\cos \theta'}{\cos \theta} = \tan \theta \sin \alpha \sin \varphi + \cos \alpha \rightarrow \cos \alpha. \tag{2.50}$$

We therefore obtain for the collision integral

$$\left. \frac{\partial f}{\partial t} \right|_{coll} = -g_1(k) \cos \theta \sum_{k'} S(k, k') \left[1 - \frac{f_0(k)g_1(k')}{f_0(k')g_1(k)} \cos \alpha \right]. \tag{2.51}$$

Note that for the relaxation approximation to be valid, the term in the brackets inside the summation sign should not depend upon the distribution function.

- Consider now the case of elastic scattering process. Then $|k| = |k'|$ and $\frac{f_0(k)g_1(k')}{f_0(k')g_1(k)} = 1$. Under these circumstances:

$$\left. \frac{\partial f}{\partial t} \right|_{coll} = -g_1(k) \cos \theta \sum_{k'} S(k, k')(1 - \cos \alpha) = -\frac{g_1(k) \cos \theta}{\tau_m(k)} = -\frac{f_A(k)}{\tau_m(k)}.$$

  Hence, when the scattering process is ELASTIC, the characteristic time τ_f equals the momentum relaxation time.

- If the scattering process is ISOTROPIC, then $S(\mathbf{k}, \mathbf{k}')$ does not depend upon α. In this case, the second term in the square brackets averages to zero and

$$\left. \frac{\partial f}{\partial t} \right|_{coll} = -g_1(k) \cos \theta \sum_{k'} S(k, k') = -\frac{g_1(k) \cos \theta}{\tau(k)} = -\frac{f_A(k)}{\tau(k)}.$$

  Thus, in this case, the characteristic time is the scattering time or the average time between collision events.

To summarize the discussion so far, under low-field conditions, when the scattering process is either isotropic or elastic, the collision term can be represented as $-f_A/\tau_f(k)$, where in general $\tau_f(k)$ is the momentum relaxation time that depends only upon the nature of the scattering process. Following these simplifications, the BTE can thus be written as

$$-eEv \cos \theta \left(\frac{\partial f_0}{\partial E} \right) = -\frac{g_1(k) \cos \theta}{\tau_m(k)} \tag{2.52}$$

or

$$g_1(k) = eEv\tau_m(k) \left(\frac{\partial f_0}{\partial E} \right). \tag{2.53}$$

The distribution function is, thus, equal to

$$f(k) = f_0(k) + eEv\cos\theta\,\tau_{\mathrm{m}}(k)\left(\frac{\partial f_0}{\partial E}\right)$$
$$= f_0(k) + eEv_z\tau_{\mathrm{m}}(k)\left(\frac{\partial f_0}{\partial E}\right).$$

(2.54)

To further investigate the form of this distribution function, we use $\frac{\partial f_0}{\partial E} = \frac{1}{\hbar v_z}\frac{\partial f_0}{\partial k_z}$. This leads to

$$f(k) = f_0(k) + \frac{e}{\hbar}E\tau_{\mathrm{m}}(k)\frac{\partial f_0}{\partial k_z}.$$

(2.55)

The second term of the last expression resembles the linear term in the Taylor series expansion of $f(\mathbf{k})$. Hence, we can write

$$f(k_x, k_y, k_z) = f_0\left(k_x, k_y, k_z + \frac{e}{\hbar}E\tau_{\mathrm{m}}(k)\right).$$

(2.56)

To summarize, the assumption made at arriving at this last result for displaced Maxwellian form for the distribution function is that the electric field $\mathbf{E}$ is small. Hence, the displaced Maxwellian is a good representation of the distribution function under low-field conditions.

PROBLEMS FOR CHAPTER 2:

1. Electrons in a lattice see a periodic potential due to the presence of the atoms, which is of the form shown below:

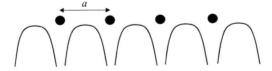

This periodic potential will open gaps in the dispersion relation, i.e., it will impose limits on the allowed particle energies. To simplify the problem, one can assume that the width of the potential energy term goes to zero, i.e., the periodic potential can be represented as an infinite series of δ-function potentials:

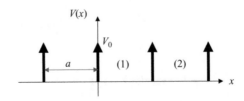

(a) Using Bloch theorem for the general form of the solution of the 1D TISE (time-independent Schrödinger equation) in the presence of periodic potential, show that the relationship between the crystal momentum and particle energy is obtained by solving the following implicit equation

$$\cos(ka) = P\frac{\sin(k_0 a)}{k_0 a} + \cos(k_0 a),$$

where k is the crystal momentum and $E = \hbar^2 k_0^2/2m$. The quantity $P = 2mV_0/\hbar^2$ in the above equation may be regarded as the "scattering power" of a single potential spike.

(b) Plot the dispersion relation for a particle in a periodic potential for the case when $P = 2.5$.

2. In the free-electron approximation, the total energy of the electrons is assumed to be always large compared to the periodic potential energy. Writing the 1D time-independent SWE in the form

$$\frac{d^2\psi}{dx^2} + \left[k_0^2 + \gamma f(x)\right]\psi(x) = 0,$$

where $E = \hbar^2 k_0^2/2m$ and $\gamma f(x) = 2mV(x)/\hbar^2$, it is rather straightforward to show that in the limit $\gamma \to 0$ and away from the band-edge points $(\pm n\pi/a)$, one can approximate the wavefunction $\psi(x)$ with

$$\psi(x) = e^{ikx}u_k(x) = e^{ikx}\sum_n b_n e^{-i2\pi nx/a} \approx b_0 e^{ikx} + \gamma e^{ikx}\sum_{n\neq 0} b_n e^{-i2\pi nx/a}.$$

(a) Show that the result given above is a valid approximation for $\psi(x)$ in the limit $\gamma \to 0$ and away from the band-edge points.

(b) Find the relationship between the expansion coefficients b_n and the Fourier expansion coefficients for the periodic potential $V(x)$ for this case. Also, obtain an analytic expression for the dispersion relation (the relationship between the allowed particle energies and the crystal momentum k) for values of the crystal momentum away from the band-edge points.

(c) How will the results from parts (a–b) change if the crystal momentum approaches the band-edge points $k_n = \pm n\pi/a$? What is the appropriate approximate expression for the wavefunction in this case? Evaluate the dispersion relation in the vicinity of the band-edge points and discuss the overall energy-wavevector dispersion relation in the free-electron approximation.

3. In the free-electron approximation, discussed in problem 2, it was assumed that the potential energy of the electron is small compared to its total energy. The atoms are assumed to be very close to each other, so that there is significant overlap between the wavefunctions for the electrons associated with neighboring atoms. This leads to wide energy bands and very narrow energy gaps. The tight-binding approximation proceeds from the opposite limit, i.e., it assumes that the potential energy of the electron accounts for nearly all of the total energy. The atoms are assumed to be very far apart so that the wavefunctions for the electrons associated with neighboring atoms overlap only to a small extent. A brief description of the tight-binding method is given below:

If the potential function associated with an isolated atom is $V_0(\mathbf{r})$, then the solution of the Schrödinger equation

$$\hat{H}_0 \psi_0(\mathbf{r}) = \left[-\frac{\hbar^2}{2m} \nabla^2 + V_0(\mathbf{r}) \right] \psi_0(\mathbf{r}) = E_0 \psi_0(\mathbf{r})$$

describes the electronic wavefunctions of the atom. If the ground-state wavefunction is not much affected by the presence of the neighboring atoms, then the crystal wavefunction is given by

$$\psi(\mathbf{r}) = \sum_n e^{i\mathbf{k}_n \cdot \mathbf{r}} \psi_0(\mathbf{r} - \mathbf{r}_n)$$

and the periodic potential is represented as

$$V(\mathbf{r}) = \sum_n V_0(\mathbf{r} - \mathbf{r}_n),$$

where $\mathbf{r}_n$ is a vector from some reference point in space to a particular lattice site. Now, if we express the potential energy term in the total Hamiltonian of the system as

$$V(\mathbf{r}) = V_0(\mathbf{r} - \mathbf{r}_n) + V(\mathbf{r}) - V_0(\mathbf{r} - \mathbf{r}_n) = V_0(\mathbf{r} - \mathbf{r}_n) + H'(\mathbf{r})$$

it is rather straightforward to show that the expectation value of the energy of the system is given by

$$E = E_0 + \frac{1}{N} \sum_m e^{-i\mathbf{k}\cdot\mathbf{r}_m} \int dV \psi_0^*(\mathbf{r} - \mathbf{r}_m)[V(\mathbf{r}) - V_0(\mathbf{r})]\psi_0(\mathbf{r}), \qquad (*)$$

where N is the number of atoms in the crystal.

(a) Complete the derivation that leads to the result given in (*).

(b) If we assume that the atomic wavefunctions are spherically symmetric (s-type), show that in the case the nearest–neighbor interaction are only taken into account, the energy of the system for simple cubic lattice is of the form

$$E = E_0 - \alpha - 2\beta \lfloor \cos(k_x a) + \cos(k_y a) + \cos(k_z a) \rfloor.$$

Identify the meaning of the terms α and β in the result given above. What is the approximate form of the dispersion relation for an electron moving in the x-direction with momentum $k_x \ll \pi/a$.

(c) Find the dispersion relation for an s-band in the tight-binding approximation for a body-centered and face-centered cubic crystal in the tight-binding approximation, considering overlap of nearest–neighbor wavefunctions only.

(d) Plot the form of the constant energy surfaces for several energies within the zone. Show that these surfaces are spherical for energies near the bottom of the band. Show that $\mathbf{n} \cdot \nabla_k E$ vanishes on the zone boundaries.

4. Derive the density of states function for 1D, 2D, and 3D (bulk) systems assuming parabolic energy bands.

## CHAPTER 3

The Drift–Diffusion Equations and Their Numerical Solution

In Chapter 1, we discussed the various levels of approximations that are employed in the modeling of semiconductor devices, and then looked at the semiclassical description of charge transport via the Boltzmann Transport Equation (BTE) in Chapter 2. However, the direct solution of the full BTE is challenging computationally, particularly when combined with field solvers for device simulation. Therefore, for traditional semiconductor device modeling, the predominant model corresponds to solutions of the so-called drift–diffusion (DD) equations, which are "local" in terms of the driving forces (electric fields and spatial gradients in the carrier density), i.e., the current at a particular point in space only depends on the instantaneous electric fields and concentration gradient at that point. The present chapter is devoted to the DD model and its application to semiconductor device modeling. We first look at the derivation of the DD model from the BTE, and the physical significance of the parameters associated with these equations. We then look at the numerical solution of the DD equations coupled with Poisson's equation in the domain of the semiconductor, leading to the Sharfetter–Gummel algorithm which is widely used in conventional device simulation.

3.1 DRIFT–DIFFUSION MODEL

The popular DD current equations can be easily derived from the BTE by considering moments of the BTE. Consider steady-state conditions and, for simplicity, a 1D geometry. With the use of a relaxation time approximation as in Eq. (2.39), the BTE may be written as [45]

$$\frac{eE}{m^*}\frac{\partial f}{\partial v} + v\frac{\partial f}{\partial v} = \frac{f_0 - f(v, x)}{\tau} \qquad (3.1)$$

In writing Eq. (3.1) parabolic bands have been assumed for simplicity, and the charge e has to be taken with the proper sign of the particle (positive for holes and negative for electrons). The general definition of current density as given in Eq. (2.19) is repeated here for completeness

$$J(x) = e\int vf(v, x)dv, \qquad (3.2)$$

where the integral on the right-hand side represents the first "moment" of the distribution function, which is discussed in detail in Chapter 4. This definition of current can be related to Eq. (3.1) after multiplying both sides of (3.1) by v and integrating over v. From the RHS of Eq. (3.1) we get

$$\frac{1}{\tau}\left[\int vf_0 dv - \int vf(v, x)dv\right] = -\frac{J(x)}{e\tau}. \tag{3.3}$$

The equilibrium distribution function is symmetric in v, and hence the first integral is zero. Therefore, we have

$$J(x) = -e\frac{e\tau}{m^*}E\int v\frac{\partial f}{\partial v}dv - e\tau\frac{d}{dx}\int v^2 f(v, x)dv. \tag{3.4}$$

Integrating by parts we have

$$\int v\frac{\partial f}{\partial v}dv = [vf(v, x)]_{-\infty}^{\infty} - \int f(v, x)dv = -n(x) \tag{3.5}$$

and we can write

$$\int v^2 f(v, x)dv = n(x)\langle v^2\rangle, \tag{3.6}$$

where $\langle v^2\rangle$ is the average of the square of the velocity. The DD equations are derived by introducing the mobility $\mu = \frac{e\tau}{m^*}$ and replacing $\langle v^2\rangle$ with its average equilibrium value $\frac{k_B T}{m^*}$ for a 1D case and $\frac{3k_B T}{m^*}$ for a 3D case, therefore neglecting thermal effects. The diffusion coefficient D is also introduced, and the resulting DD current expressions for electrons and holes are

$$J_n = qn(x)\mu_n E(x) + qD_n\frac{dn}{dx},$$
$$J_p = qp(x)\mu_p E(x) - qD_p\frac{dp}{dx}, \tag{3.7}$$

respectively, where q is used to indicate the absolute value of the electronic charge. Although no direct assumptions on the nonequilibrium distribution function, $f(v,x)$, were made in the derivation of Eqs. (3.7), in effect, the choice of equilibrium (thermal) velocity means that the DD equations are only valid for small perturbations from the equilibrium state (low fields). The validity of the DD equations is empirically extended by introduction of a field-dependent mobility $\mu(E)$ and diffusion coefficient $D(\mathbf{E})$, obtained from empirical models or detailed calculation to capture effects such as velocity saturation at high electric fields due to hot carrier effects. A more detailed description of the most commonly used models for the carrier mobility is given in Appendix B.

3.1.1   Physical Limitations on Numerical Drift–Diffusion Schemes

The complete DD model is based on the following set of equations in 1D:

1. Current equations

$$J_n = q\,n(x)\mu_n E(x) + q\,D_n\frac{dn}{dx},$$

$$J_p = q\,p(x)\mu_p E(x) - q\,D_p\frac{dp}{dx}. \qquad (3.8)$$

2. Continuity equations

$$\frac{\partial n}{\partial t} = \frac{1}{q}\nabla\cdot\mathbf{J}_n + U_n,$$

$$\frac{\partial p}{\partial t} = -\frac{1}{q}\nabla\cdot\mathbf{J}_p + U_p. \qquad (3.9)$$

3. Poisson's equation

$$\nabla\cdot\varepsilon\nabla V = -(p - n + N_p^+ - N_A^-), \qquad (3.10)$$

where U_n and U_p are the net generation–recombination rates.

   The continuity equations are the conservation laws for the charge carriers, which may be easily derived taking the zeroth moment of the time dependent BTE as discussed in detail later in Chapter 4. A numerical scheme which solves the continuity equations should

1. Conserve the total charge inside the device, as well as that entering and leaving.

2. Respect the local positive definite nature of carrier density. Negative density is unphysical.

3. Respect monotonicity of the solution (i.e., it should not introduce spurious space oscillations).

Conservative schemes are usually achieved by subdivision of the computational domain into patches (boxes) surrounding the mesh points. The currents are then defined on the boundaries of these elements, thus enforcing conservation (the current exiting one element side is exactly equal to the current entering the neighboring element through the side in common). For example, on a uniform 2D grid with mesh size Δ, the electron continuity equation may be discretized in an explicit form as follows [46]:

$$\frac{n(i, j, k + 1) - n(i, j, k)}{\Delta t} = \frac{J^x(i + \frac{1}{2}, j; k) - J^x(i - \frac{1}{2}, j; k)}{q\Delta}$$

$$+ \frac{J^y(i, j + \frac{1}{2}; k) - J^x(i, j - \frac{1}{2}; k)}{q\Delta}. \qquad (3.11)$$

In Eq. (3.11), the indices i, j describe spatial discretization, k corresponds to the time progression, and the superscripts x and y denote the x- and y-coordinate of the current density vector.

This simple approach has certain practical limitations, but is sufficient to illustrate the idea behind the conservative scheme. With the present convention for positive and negative components, it is easy to see that in the absence of generation–recombination terms, the only contributions to the overall device current arise from the contacts. Remember that, since electrons have negative charge, the particle flux is opposite to the current flux. The actual determination of the current densities appearing in Eq. (3.11) will be discussed later. When the equations are discretized, using finite differences for instance, there are limitations on the choice of mesh size and time step [47]:

1. The mesh size Δx is limited by the Debye length.
2. The time step is limited by the dielectric relaxation time.

The mesh size must be smaller than the Debye length where one has to resolve charge variations in space. A simple example is the carrier redistribution at an interface between two regions with different doping levels. Carriers diffuse into the lower doped region creating excess carrier distribution which at equilibrium decays in space down to the bulk concentration with approximately exponential behavior. The spatial decay constant is the Debye length

$$L_D = \sqrt{\frac{\varepsilon k_B T}{q^2 N}}, \tag{3.12}$$

where N is the doping density. In GaAs and Si, at room temperature the Debye length is approximately 400 Å when $N \approx 10^{16} \text{cm}^{-3}$ and decreases to about 50 Å when $N \approx 10^{18} \text{cm}^{-3}$.

The dielectric relaxation time is the characteristic time for charge fluctuations to decay under the influence of the field that they produce. The dielectric relaxation time may be estimated using

$$t_{dr} = \frac{\varepsilon}{q N \mu}. \tag{3.13}$$

To see the importance of respecting the limitations related to the dielectric relaxation time, imagine we have a spatial fluctuation of the carrier concentration, which relaxes to equilibrium according to the rate equation

$$\frac{\partial \Delta n}{\partial t} = \frac{\Delta n(t = 0)}{t_{dr}}. \tag{3.14}$$

A finite-difference discretization of this equation gives at the first time step

$$\Delta n(\Delta t) = \Delta n(0) - \frac{\Delta t \Delta n(0)}{t_{dr}}. \tag{3.15}$$

Clearly, if $\Delta t > t_{dr}$, the value of $\Delta n(\Delta t)$ is negative, which means that the actual concentration is evaluated to be less than the equilibrium value, and it is easy to see that the solution at higher time steps will decay oscillating between positive and negative values of Δn. An excessively large Δt may lead, therefore, to nonphysical results. In the case of high mobility, the dielectric relaxation time can be very small. For instance, a sample of GaAs with a mobility of $6000\,\mathrm{cm}^2(V\,s)^{-1}$ and doping $10^{18}\mathrm{cm}^{-3}$ has approximately $t_{\mathrm{dr}} \approx 10^{-15}$ s, and in a simulation the time step Δt should be chosen to be considerably smaller.

3.1.2   Steady-State Solution of Bipolar Semiconductor Equations

The general semiconductor equations for electrons and holes may be rewritten in 3D as

$$\nabla.(\varepsilon \nabla V) = q(n - p + N_{\mathrm{B}}),$$

$$\nabla.J_n = qU(n,\,p) + q\frac{\partial n}{\partial t},$$

$$\nabla.J_p = qU(n,\,p) + q\frac{\partial p}{\partial t},$$

$$J_n = -q\mu_n\left(-n\nabla V + \frac{k_{\mathrm{B}}T}{q}\nabla n\right),$$

$$J_p = -q\mu_p\left(-p\nabla V + \frac{k_{\mathrm{B}}T}{q}\nabla p\right), \tag{3.16}$$

with $N_{\mathrm{B}} = N_{\mathrm{A}} - N_{\mathrm{D}}$. We note that the above equations are valid in the limit of small deviations from equilibrium, since the Einstein relations have been used for the diffusion coefficient, normally valid for low fields or large devices. The generation–recombination term U will be in general a function of the local electron and hole concentrations, according to possible different physical mechanisms, to be examined later in more detail. We will consider from now on steady state, with the time dependent derivatives set to zero.

These semiconductor equations constitute a coupled nonlinear set. It is not possible, in general, to obtain a solution directly in one step, rather a nonlinear iteration method is required. The two more popular methods for solving the discretized equations are the Gummel's iteration method [48] and the Newton's method [49]. It is very difficult to determine an optimum strategy for the solution, since this will depend on a number of details related to the particular device under study. There are in general three possible choices of variables.

1. Natural variable formulation (V, n, p)

2. Quasi-Fermi level formulation V, ϕ_n, ϕ_p, where the quasi-Fermi levels derive from the following definition of carrier concentration out of equilibrium (for nondegenerate case)

$$n = n_i \exp\left(\frac{q(V - \phi_n)}{k_{\mathrm{B}}T}\right),$$

$$n = n_i \exp\left(\frac{q(\phi_p - V)}{k_{\mathrm{B}}T}\right).$$

3. Slotboom formulation (V, Φ_n, Φ_p) where the Slotboom [50] variables are defined as

$$\Phi_n = n_i \exp\left(-\frac{q\phi_n}{k_B T}\right).$$

$$\Phi_p = n_i \exp\left(\frac{q\phi_p}{k_B T}\right).$$

The Slotboom variables are, therefore, related to the carrier density expressions, and the extension to degenerate conditions is cumbersome.

Normally, there is a preference for the quasi-Fermi level formulation in steady-state simulation, and for the natural variables n and p in transient simulation.

3.1.3   Normalization and Scaling

For the sake of clarity, all formulae have been presented without the use of simplifications or normalization. It is however common practice to perform the actual calculation using normalized units to make the algorithms more efficient, and in cases to avoid numerical overflow and underflow. It is advisable to input the data in M.K.S. or practical units (the use of centimeters is for instance very common in semiconductor practice, instead of meters) and then provide a conversion block before and after the computation blocks to normalize and denormalize the variables. It is advisable to use consistent scaling, rather than set certain constants to arbitrary values. The most common scaling factors for normalization of semiconductor equations are listed in Table 3.1 [51].

3.1.4   Gummel's Iteration Method

Gummel's method [46], [48] solves the coupled set of semiconductor equations (Eqs. (3.16)) together with the Poisson equation via a decoupled procedure. If we choose the quasi-Fermi level formulation, we solve first a nonlinear Poisson's equation. The potential obtained from this solution is substituted into the continuity equations, which are now linear, and are solved directly to conclude the iteration step. The result in terms of quasi-Fermi levels is then substituted back into Poisson's equation and the process repeated until convergence is reached. In order to check for convergence, one can calculate the residuals obtained by positioning all the terms to the left-hand side of the equations and substituting the variables with the iteration values. For the exact solution the residuals should be zero. Convergence is assumed when the residuals are smaller than a set tolerance. The rate of convergence of the Gummel method is faster when there is little coupling between the different equations. The computational cost of one Gummel iteration is one matrix solution for each carrier type plus one iterative solution for the linearization of Poisson's equation. Note that in conditions of equilibrium (zero bias) only the solution of Poisson's equation is necessary, since the equilibrium Fermi level is constant and coincides with both quasi-Fermi levels.

TABLE 3.1: Scaling factors

VARIABLE	SCALING VARIABLE	FORMULA
Space	Intrinsic Debye length ($N = n_i$) Extrinsic Debye length ($N = N_{max}$)	$L = \sqrt{\frac{\varepsilon k_B T}{q^2 N}}$
Potential	Thermal voltage	$V* = \frac{k_B T}{q}$
Carrier concentration	Intrinsic concentration Maximum doping concentration	$N = n_i$ $N = N_{max}$
Diffusion coefficient	Practical unit Maximum diffusion coefficient	$D = 1\, cm^2 s^{-1}$ $D = D_{max}$
Mobility		$M = \frac{D}{V*}$
Generation–recombination		$R = \frac{DN}{L^2}$
Time		$T = \frac{L^2}{D}$

We give some examples of the quasi-linearization of Poisson equation, as necessary when Gummel's method is implemented. We start with a 1D case example, and then we give two 2D examples, one for discretization of a bulk material and the second one for discretization of a 2D device at the semiconductor oxide interface. Let us consider the 1D case in equilibrium first. As mentioned earlier, one has to solve only Poisson's equation, since the current is zero and the exact expressions for the carrier concentrations are known. In the nondegenerate case, the explicit expressions for the electron and hole densities are substituted into Poisson's equation to give

$$\frac{d^2 V}{dx^2} = \frac{q}{\varepsilon}\left[n_i \exp(-q\phi_n)\exp\left(\frac{qV}{k_B T}\right) - n_i \exp(q\phi_p)\exp\left(-\frac{qV}{k_B T}\right) + N_A - N_D \right], \quad (3.17)$$

which is sometimes referred to as the nonlinear Poisson equation due to the nonlinear terms involving V on the RHS. In equilibrium, the quasi-Fermi energies are equal and spatially invariant, hence we may choose the Fermi energy as the reference energy, i.e., $\phi_n = \phi_p = 0$. Furthermore, the equation may be scaled by using the (minimum) extrinsic Debye length for the space coordinate x, and the thermal voltage $k_B T/q$ for the potential V. Writing $\bar{V}$ and $\bar{x}$ for the normalized potential and space coordinates, we obtain

$$\frac{d^2 \bar{V}}{d\bar{x}^2} = \frac{n_i}{N}\left[\exp(\bar{V}) - \exp(-\bar{V}) + \frac{N_A - N_D}{n_i} \right]. \quad (3.18)$$

The equilibrium nonlinear Poisson equation can be solved with the following quasi-linearization procedure

1. Choose an initial guess for the potential $\bar{V}$.

2. Write the potential at the next iteration step as $\bar{V}_{\text{new}} = \bar{V} + \delta V$, and substitute into Eq. (3.18) to solve for $\bar{V}_{\text{new}}$ to give

$$\frac{d^2\bar{V}}{d\bar{x}^2} + \frac{d^2\delta V}{d\bar{x}^2} = \frac{n_i}{N}\left[\exp(\bar{V})\exp(\delta V) - \exp(-\bar{V})\exp(-\delta V) + \frac{N_A - N_D}{n_i}\right]. \quad (3.19)$$

3. Use the linearization $\exp(\pm\delta V) \approx 1 \pm \delta V$ and discretize the resultant equation. This equation has a tridiagonal matrix form and is readily solved for $\delta V(i)$.

$$\delta V(i-1) - \left[2 + \frac{n_i}{N}\Delta^2 x[\exp(\bar{V}(i)) - \exp(-\bar{V}(i))]\right]\delta V(i) + \delta V(i+1)$$
$$= -\bar{V}(i-1) + 2\bar{V}(i) - \bar{V}(i+1) + \frac{n_i}{N}\Delta^2 x\left[\exp(\bar{V}(i)) - \exp(-\bar{V}(i)) + \frac{N_A - N_D}{n_i}\right].$$
$$(3.20)$$

4. Check for convergence. The residual of Eq. (3.20) is calculated and convergence is achieved if the norm of the residual is smaller than a preset tolerance. If convergence is not achieved, return to step 2. In practice one might simply check the norm of the error

$$||\delta V||_2 \leq \text{Tol} \quad \text{or} \quad ||\delta V||_\infty \geq \text{Tol}.$$

Note that for the solution of the nonlinear Poisson's equation, the boundary conditions are referenced to the equilibrium Fermi level. One may use the separation between the Fermi level and the intrinsic Fermi level at the contacts for the boundary conditions.

After the solution in equilibrium is obtained, the applied voltage is increased gradually in steps of $\Delta V \leq k_B T/q$ to avoid numerical instability. The scaled nonlinear Poisson equation under nonequilibrium conditions now becomes

$$\frac{d^2 V}{dx^2} = \frac{n_i}{N}\left[\exp(-\phi_n)\exp(V) - \exp(-\phi_p)\exp(-V) + \frac{N_A - N_D}{n_i}\right], \quad (3.21)$$

where the quasi-Fermi levels are also normalized. Assuming Einstein's relations still hold, the current density equation may be re-written as

$$J_n = -q\mu_n n\frac{\delta V}{\delta x} + q\mu_n\frac{k_B T}{q}\frac{\partial}{\partial x}\left[n_i\exp\left(\frac{q(V - \phi_n)}{k_B T}\right)\right]$$
$$= -q\mu_n n\frac{\delta V}{\delta x} + q\mu_n\frac{k_B T}{q}n\frac{\partial}{\partial_B x}\left[\frac{\partial V}{\partial x} - \frac{\partial\phi_n}{\partial x}\right]$$
$$= -q\mu_n n\frac{\delta\phi_n}{\delta x}$$

$$= -q\mu_n n_i \exp\left[\frac{q(V-\phi_n)}{k_B T}\right]\frac{\delta\phi_n}{\delta x}$$

$$= -q\mu_n n_i \exp\left(\frac{qV}{k_B T}\right)\frac{-k_B T}{q}\frac{\partial}{\delta x}\exp\left(\frac{-q\phi_n}{k_B T}\right), \tag{3.22}$$

which may be written more compactly, including quasi-Fermi level normalization, as

$$J_n = a_n(x)\frac{\delta}{\delta x}\exp(-\phi_n). \tag{3.23}$$

A similar formula is obtained for the holes

$$J_p = a_p(x)\frac{\delta}{\delta x}\exp(\phi_p) \tag{3.24}$$

and the continuity equations are therefore given by

$$\frac{\delta}{\delta x}\left[a_n(x)\frac{\delta}{\delta x}\exp(-\phi_n)\right] = qU(x), \tag{3.25}$$

$$\frac{\delta}{\delta x}\left[a_p(x)\frac{\delta}{\delta x}\exp(\phi_p)\right] = qU(x). \tag{3.26}$$

The continuity equations may be discretized with a straightforward finite-difference approach (here for simplicity with uniform mesh)

$$\frac{\dfrac{a_\alpha\left(i+\frac{1}{2}\right)[\Phi_\alpha(i+1)-\Phi_\alpha(i)]}{\Delta x} - \dfrac{a_\alpha\left(i-\frac{1}{2}\right)[\Phi_\alpha(i)-\Phi_\alpha(i-1)]}{\Delta x}}{\Delta x} = u, \tag{3.27}$$

where the Slotboom variables have been used for simplicity of notation. Note that the inner derivative has been discretized with centered differences around the points $\left(i\pm\frac{1}{2}\right)$ of the interleaved mesh. Variables on the interleaved mesh must be determined very carefully, using consistent interpolation schemes for potential and carrier density, as discussed later. The discretized continuity equations lead to the tridiagonal system

$$a_n\left(i-\frac{1}{2}\right)\Phi_n(i-1) - \left[a_n\left(i+\frac{1}{2}\right) + a_n\left(i-\frac{1}{2}\right)\right]\Phi_n(i)$$
$$+ a_n\left(i+\frac{1}{2}\right)\Phi_n(i+1) = \Delta^2 x U(i), \tag{3.28}$$

$$a_p\left(i-\frac{1}{2}\right)\Phi_p(i-1) - \left[a_p\left(i+\frac{1}{2}\right) + a_p\left(i-\frac{1}{2}\right)\right]\Phi_p(i)$$
$$+ a_p\left(i+\frac{1}{2}\right)\Phi_p(i+1) = -\Delta^2 x U(i). \tag{3.29}$$

The decoupled iteration now solves Poisson's Eq. (3.18), initially with a guess for the quasi-Fermi levels. The voltage distribution obtained for the previous voltage considered is normally a good initial guess for the potential. Since the quasi-Fermi levels are inputs for Poisson's equation, the quasi-linearization procedure for equilibrium can be used again. The potential is then used to update the $a_n(i)$ and $a_p(i)$, Eqs. (3.28) and (3.29) are solved to provide new quasi-Fermi level values for Poisson's equation, and the process is repeated until convergence is reached. The generation–recombination term depends on the electron and hole concentrations, therefore it has to be updated at each iteration. It is possible to update the generation–recombination term also intermediately, using the result of Eq. (3.28) for the electron concentration.

The examples given bellow illustrates the Gummel's approach and is limited to the non-degenerate case. If field dependent mobility and diffusion coefficients are introduced, minimal changes should be necessary, as long as it is still justified the use of Einstein's relations. Extension to a nonuniform mesh is left as an exercise for the reader. In the 2D case, the quasi-linearized Poisson's equation becomes

$$
-\left(4 + h^2 \frac{n_i}{N}[\Phi_n(i, j)\exp(V(i, j)) + \Phi_p(i, j)\exp(-V(i, j))]\right)\partial V(i, j)
$$
$$
+ [\partial V(i-1, j) + \partial V(i+1, j) + \partial V(i, j-1) + \partial V(i, j+1)]
$$
$$
= 4V(i, j) - V(i-1, j) - V(i+1, j) - V(i, j-1) - V(i, j+1) + h^2 \frac{n_i}{N}
$$
$$
\left[\Phi_n(i, j)\exp(V(i, j)) + \Phi_p(i, j)\exp(-V(i, j)) + \frac{N_A + N_B}{N_i}\right]. \tag{3.30}
$$

The normalized mesh size is $h = \Delta x = \Delta y$. As before, the thermal voltage $k_B T/q$ has been used to normalize the potential V and the quasi-Fermi levels ϕ_n and ϕ_p included in the Slotboom variables $\Phi_{n,p} = \exp(\pm\phi_{n,p})$.

The continuity equations with the form $\nabla.(a(x, y)\nabla\Phi) = \pm U(x, y)$ are discretized as

$$
-\left[a\left(1 + \frac{1}{2}, j\right) + a\left(1 - \frac{1}{2}, j\right) + a\left(i, j + \frac{1}{2}\right) + a\left(i, j - \frac{1}{2}\right)\right]\Phi(i, j)
$$
$$
+ a\left(i + \frac{1}{2}, j\right)\Phi(i+1, j) + a\left(i - \frac{1}{2}, j\right)\Phi(i-1, j) + a\left(i, j + \frac{1}{2}\right)\Phi(i, j+1)
$$
$$
+ a\left(i, j - \frac{1}{2}\right)\Phi(i, j-1) = \pm h^2 U(i, j). \tag{3.31}
$$

3.1.5   Newton's Method

Newton's method is a coupled procedure which solves the equations simultaneously, through a generalization of the Newton–Raphson method for determining the roots of an equation. We rewrite Eqs. (3.8–3.10) in the residual form

$$
W_v(V, n, p) = 0 \quad W_n(V, n, p) = 0 \quad W_p(V, n, p) = 0. \tag{3.32}
$$

Starting from an initial guess V_0, n_0, and p_0, the corrections V, Δn, and Δp are calculated from the Jacobian system

$$\begin{pmatrix} \dfrac{\delta W_v}{\delta V} & \dfrac{\delta W_v}{\delta n} & \dfrac{\delta W_v}{\delta p} \\[8pt] \dfrac{\delta W_n}{\delta V} & \dfrac{\delta W_n}{\delta n} & \dfrac{\delta W_n}{\delta p} \\[8pt] \dfrac{\delta W_p}{\delta V} & \dfrac{\delta W_p}{\delta n} & \dfrac{\delta W_p}{\delta p} \end{pmatrix} \begin{pmatrix} \Delta V \\[8pt] \Delta n \\[8pt] \Delta p \end{pmatrix} = - \begin{pmatrix} W_V \\[8pt] W_n \\[8pt] W_p \end{pmatrix}, \tag{3.33}$$

which is obtained by Taylor expansion. The solutions are then updated according to the scheme

$$\begin{aligned} V(K+1) &= V(k) + \Delta V(k) \\ n(K+1) &= n(k) + \Delta n(k) \\ p(K+1) &= p(k) + \Delta p(k), \end{aligned} \tag{3.34}$$

where k indicates the iteration number. In practice, a relaxation approach is also applied to avoid excessive variations of the solutions at each iteration step.

The system (3.34) has three equations for each mesh point on the grid. This indicates the main disadvantage of a full Newton iteration, related to the computational cost of matrix inversion (one may estimate that a $3N \times 3N$ matrix takes typically 20 times longer to invert than an analogous $N \times N$ matrix). On the other hand convergence is usually fast for the Newton method, provided that the initial condition is reasonably close to the solution, and is in the neighborhood where the solution is unique. There are several viable approaches to alleviate the computational requirements of the Newton's method. In the Newton–Richardson approach, the Jacobian matrix in Eq. (3.33) is updated only when the norm of the error does not decrease according to a preset criterion. In general, the Jacobian matrix is not symmetric positive definite, and fairly expensive solvers are necessary. Iterative schemes have been proposed to solve each step of Newton's method by reformulating Eq. (3.33) as

$$\begin{pmatrix} \dfrac{\delta W_v}{\delta V} & 0 & 0 \\[8pt] \dfrac{\delta W_n}{\delta V} & \dfrac{\delta W_n}{\delta n} & 0 \\[8pt] \dfrac{\delta W_p}{\delta V} & \dfrac{\delta W_p}{\delta n} & \dfrac{\delta W_p}{\delta p} \end{pmatrix} \begin{pmatrix} \Delta V \\[8pt] \Delta n \\[8pt] \Delta p \end{pmatrix}_{k+1} = - \begin{pmatrix} W_V \\[8pt] W_n \\[8pt] W_p \end{pmatrix} - \begin{pmatrix} 0 & \dfrac{\delta W_v}{\delta n} & \dfrac{\delta W_v}{\delta p} \\[8pt] 0 & 0 & \dfrac{\delta W_n}{\delta p} \\[8pt] 0 & 0 & 0 \end{pmatrix} \begin{pmatrix} \Delta V \\[8pt] \Delta n \\[8pt] \Delta p \end{pmatrix}_{k}. \tag{3.35}$$

Since the matrix on the left-hand side is lower triangular, one may solve Eq. (3.35) by decoupling into three systems of equations solved in sequence. First, one solves the block of equations (again, one for each grid point)

$$\frac{\delta W_v}{\delta V}(\Delta V)_{k+1} = -W_V - \frac{\delta W_v}{\delta n}(\Delta n)_k - \frac{\delta W_v}{\delta p}(\Delta P)_k \tag{3.36}$$

and the result is used in the next block of equations

$$\frac{\delta W_n}{\delta n}(\Delta n)_{k+1} = -W_n - \frac{\delta W_n}{\delta V}(\Delta V)_{k+1} - \frac{\delta W_n}{\delta p}(\Delta p)_k. \qquad (3.37)$$

Similarly, for the third block

$$\frac{\delta W_p}{\delta p}(\Delta p)_{k+1} = -W_p - \frac{\delta W_p}{\delta V}(\Delta V)_{k+1} - \frac{\delta W_p}{\delta n}(\Delta n)_{k+1}. \qquad (3.38)$$

The procedure achieves a decoupling of the equations as in a block Gauss-Seidel iteration, and can be intended as a generalization of the Gummel method. A block Successive Over Relaxation (SOR) method is obtained if the left-hand sides are premultiplied by a relaxation parameter. This iteration procedure has better performance if the actual variables are (V, ϕ_n, ϕ_p).

In general, Gummel's method is preferred at low bias because of its faster convergence and low cost per iteration. At medium and high bias the Newton's method becomes more convenient, since the convergence rate of Gummel's method becomes worse as the coupling between equations becomes stronger at higher bias. But since Gummel's method has a fast initial error reduction, it is often convenient to couple the two procedures, using Newton's method after several Gummel's iterations. Remember that it is very important for the Newton's iteration to start as close as possible to the true solution. Close to convergence, the residual in Newton's iteration should decrease quadratically from one iteration to the other.

3.1.6   Generation and Recombination

The Shockley–Read–Hall model is very often used for the generation–recombination term due to trap levels

$$U_{\text{SRH}} = \frac{np - n_i^2}{\tau_p \left[n + n_i \exp\left(\frac{q(E_t - E_i)}{k_B T} \right) \right] + \tau_n \left[p + n_i \exp\left(\frac{q(E_i - E_t)}{k_B} \right) \right]}, \qquad (3.39)$$

where E_t is the trap energy level involved and τ_n and τ_p are the electron and hole lifetimes. Surface states may be included with a similar formula, in which the lifetimes are substituted by $1/S_{n,p}$ where $S_{n,p}$ is the surface recombination velocity.

Auger recombination may be accounted for by using the formula

$$U_{\text{Aug}} = C_n[pn^2 - nn_i^2] + C_p[np^2 - pn_i^2], \qquad (3.40)$$

where C_n and C_p are appropriate constants. The Auger effect is for instance very relevant in the modeling of highly doped emitter regions in bipolar transistors.

The generation rate due to impact ionization can be included using the field-dependent rate

$$U_I = \frac{a_n^\infty \exp\left(\frac{-E_n^{crit}}{E}\right)^{\beta n} |J_n| + a_p^\infty \exp\left(\frac{-E_p^{crit}}{E}\right)^{\beta p} |J_p|}{q}, \qquad (3.41)$$

where E_n^{crit} and E_p^{crit} are the critical electrical fields for the onset of impact ionization initiated by electrons and holes respectively.

3.1.7 Time-Dependent Simulation

The time-dependent form of the DD equations can be used both for steady-state and transient calculations. Steady-state analysis is accomplished by starting from an initial guess, and letting the numerical system evolve until a stationary solution is reached, within set tolerance limits. This approach is seldom used in practice, since now robust steady-state simulators are widely available. It is nonetheless an appealing technique for beginners since a relatively small effort is necessary for simple applications and elementary discretization approaches. If an explicit scheme is selected, no matrix solutions are necessary, but it is normally the case that stability is possible only for extremely small time-steps.

The simulation of transients requires the knowledge of a physically meaningful initial condition, which can be obtained from a steady-state calculation. The same time-dependent numerical approaches used for steady-state simulation are suitable, but there must be more care for the boundary conditions, because of the presence of displacement current during transients. In a transient simulation to determine the steady-state, the displacement current can be neglected because it goes to zero when a stationary condition is reached. Therefore, it is sufficient to impose on the contacts the appropriate potential values provided by the bias network. In a true transient regime, however, the presence of displacement currents manifests itself as a potential variation at the contacts, superimposed on the bias, which depends on the external circuit in communication with the contacts. Neglect of the displacement current in a transient is equivalent to the application of bias voltages using ideal voltage generators, with zero internal impedance. In such a situation, the potential variations due to displacement current drop across a short circuit, and are therefore cancelled. In this arrangement, one will observe the shortest possible switching time attainable with the structure considered, but in practice an external load and parasitics will be present, and the switching times will be normally longer. A simulation neglecting displacement current effects may be useful to assess the ultimate speed limits of a device structure.

When a realistic situation is considered, it is necessary to include a displacement term in the current equations. It is particularly simple to deal with a 1D situation. Consider a 1D device with length W and a cross-sectional area A. The total current flowing in the device is

$$I_D(t) = I_n(x, t) + cA\frac{\partial E(x, t)}{\partial t}. \qquad (3.42)$$

The displacement term makes the total current constant at each position x. This property can be exploited to perform an integration along the device

$$I_D(t) = \frac{1}{W} \int_0^w I_n(x, t)dx + \frac{cA}{W}\frac{\partial V^*}{\partial t}, \qquad (3.43)$$

where $V^*(t)$ is the total voltage drop across the structure, with the ground reference voltage applied at $x = W$. The term $\varepsilon A/W$ is called the cold capacitance. The 1D device, therefore, can be studied as the parallel combination of a current generator and of the cold capacitance which is in parallel with the (linear) load circuit. At every time step, V^* has to be updated, since it depends on the charge stored by the capacitors.

To illustrate the procedure, consider a simple Gunn diode in parallel with an RLC resonant load containing the bias source. Calling C_0 the parallel combination of the cold and load capacitance,

$$I(t) = C_0\frac{\partial V^*}{\partial t} + I_0(t), \qquad (3.44)$$

where $I_0(t)$ is the particle current given by the first term on the right-hand side of Eq. (3.43), calculated at a given time step with DD (or any other suitable scheme). It is also

$$I(t) = \frac{V^*(t)}{R} - \int \frac{V^*(t) - V_b}{L}dt. \qquad (3.45)$$

Upon time differencing this last equation, with the use of finite differences we obtain

$$V^*(t + \Delta t) = V^*(t) + [I(t) - I_0(t)]\frac{\Delta t}{C_0} \qquad (3.46)$$

$$I(t + \Delta t) = I(t) - \frac{V^*(t + \Delta t) - V^*(t)}{R} - [V^*(t) - V_b)]\frac{\Delta t}{L}. \qquad (3.47)$$

This set of difference equations allows one to update the boundary conditions for Poisson's equation at every time step to fully include displacement current.

A robust approach for transient simulation should be based on the same numerical apparatus established for purely steady-state models. It is usually preferred to use fully implicit schemes, which require a matrix solution at each iteration, because the choice of the time-step is more likely to be limited by the physical time constants of the problem rather than by stability of the numerical scheme (the Courant–Friedrichs–Lewy condition, C.F.L.). In order to estimate the time-step limits, let us assume a typical electron velocity $v = 10^7$ cm^{-1}s and a spatial mesh $\Delta x = 0.01$ μm. The C.F.L. condition necessary to resolve correctly a purely drift process on this mesh requires $\Delta t \leq \Delta x/v = 10^{-15}$s. As calculated earlier, this value is not too far from typical values of the dielectric relaxation time in practical semiconductor structures.

When dealing with unipolar devices, as often used in many microwave applications, it is possible to formulate very simple time-dependent DD models, which can be solved with straightforward finite-difference techniques and are suitable for small student projects. If we can neglect the generation–recombination effects, the 1D unipolar DD model is reduced to the following system of equations

$$\frac{\delta n}{\delta t} = -\frac{d}{dx}[n v_{\mathrm{d}}(E)] + \frac{d}{dx}\left[D(E)\frac{d}{dx}n\right],$$ (3.48)

$$\frac{d^2 V}{dx^2} = \frac{q(n - N_{\mathrm{D}})}{\varepsilon},$$ (3.49)

where $v_{\mathrm{d}}(E) = -\mu_n(E)E$ is the drift velocity. There are two physical processes involved: drift (advection) expressed by the first term on the right-hand side of Eq. (3.48), and diffusion described by the second term. The continuity Eq. (3.48) is an admixture of competing hyperbolic and parabolic behavior whose relative importance depends on the local electric field strength. The system (3.48) and (3.49) can be used for both transient or steady-state conditions if the simulation is run until $\delta n/\delta t = 0$. A basic simple algorithm consists of the following steps

1. Guess the carrier distribution $n(x)$.

2. Solve Poisson's equation to obtain the field distribution.

3. Compute one iteration of the discretized continuity equation with time step Δt. $v(E)$ and $D(E)$ are updated according to the local field value.

4. Check for convergence. If convergence is obtained, stop. Otherwise, go back to step (2) updating the charge distribution.

This is an uncoupled procedure, since Eqs. (3.48) and (3.49) are not solved simultaneously. Usually, explicit methods are used for computational speed. The time step must respect the limitations due to the C.F.L. condition (related to the advective component) and to the dielectric relaxation time. A simple discretization scheme could employ an explicit finite-difference approach

$$n(i; k + 1) = n(i; k) + \frac{\Delta t}{\Delta x}\Big\{ [v_{\mathrm{d}}(i - 1; k)n(i - 1; k) - v_{\mathrm{d}}(i; k)n(i; k)]$$
$$+ \frac{1}{\Delta x}D(i; k)[n(i - 1; k) - 2n(i; k) + n(i + 1; k)]\Big\}; v_{\mathrm{d}} < 0,$$

$$n(i; k + 1) = n(i; k) + \frac{\Delta t}{\Delta x}\Big\{ [v_{\mathrm{d}}(i; k)n(i; k) - v_{\mathrm{d}}(i + 1; k)n(i + 1; k)]$$
$$+ \frac{1}{\Delta x}D(i; k)[n(i - 1; k) - 2n(i; k) + n(i + 1; k)]\Big\}; v_{\mathrm{d}} > 0,$$

(3.50)

where we have introduced upwinding for the drift term and we have assumed that the diffusion coefficient is slowly varying in space. There are of course many other possible explicit and implicit discretizations. Such simple finite-difference approaches are in general a compromise, which cannot provide at one time an optimal treatment of both advective and diffusive components. Because of spatially varying drift velocity, spurious diffusion and dispersion are present. This could be mitigated by using a nonuniform grid discretization, where the mesh size is locally adapted to achieve $v_d = \Delta x / \Delta t$ everywhere, which would involve interpolation to the new gridpoints. The discretization for a diffusive process is better behaved with a fully implicit scheme (if the Crank–Nicholson approach is used, one needs to make sure that spurious oscillations in the solution do not develop). On the other hand, the fully implicit algorithm for advection is not conservative. From these conflicting requirements, it emerges that it would be beneficial to split the drift and diffusion processes, and apply an optimal solution procedure to each. There are 1D situations where this is known to be nearly exact. In well-known experiments, a small concentration of excess carriers is generated in a semiconductor sample with a uniform electric field, and the motion of the centroid of the carrier envelope can be studied independently of the diffusive spread of the spatial distribution around the centroid itself. For an initial Gaussian distribution in space, a simple analytical solution shows that drift and diffusion can be treated as a sequential process, each using the total duration of the observation as simulation time. In analogy with this, the 1D continuity equation can be solved in two steps, for instance

$$n^*(j, i+1) = n(j, i) + v_d(j)[n(j-1, i) - n(j, i)]\frac{\Delta t}{\Delta x}; v_d < 0$$

$$n(j, i+1) = n^*(j, i+1) + D(j)[n(j-1, i+1) - 2n(j, i-1) + n(j+1, i+1)]\frac{\Delta t}{\Delta^2 x},$$

$$(3.51)$$

where again a simple explicit upwinding scheme is used for the drift, while a fully implicit scheme is used for the diffusion.

3.1.8   Scharfetter–Gummel Approximation

The discretization of the continuity equations in conservative form requires the determination of the currents on the mid-points of mesh lines connecting neighboring grid nodes. Since the solutions are accessible only on the grid nodes, interpolation schemes are needed to determine the currents. For consistency with Poisson's equation, it is common to assume that the potential varies linearly between two neighboring nodes. This is equivalent to assuming a constant field along the mesh lines, and the field at the mid-point is obtained by centered finite differences of the potential values. In order to evaluate the current, it is also necessary to estimate the carrier density at the mid-points. The simplest approximation which comes to mind is to also assume a linear variation of the carrier density, by taking the arithmetic average between two neighboring nodes. This simple approach is only acceptable for very small potential variation between the nodes, and indeed is exact only if the field between two nodes is zero, which implies the same carrier density on the two points.

In order to illustrate this, let us consider a 1D mesh where we want to discretize the electron current

$$J_n = q \mu_n n \left(-\frac{d\Psi}{dx}\right) + q D_n \frac{dn}{dx}. \tag{3.52}$$

Here, the field is explicitly expressed by the derivative of the potential. The discretization on the mid-point of the mesh line between nodes x_i and x_{i+1} is given by

$$J_{i+\frac{1}{2}} = -q \mu_n n_{i+1/2} \frac{\Psi_{i+1} - \Psi_i}{\Delta x} + q D_n \frac{n_{i+1} - n_i}{\Delta x}. \tag{3.53}$$

In the simple approach indicated above, the carrier density is expressed as

$$n_{i+1/2} \approx \frac{n_{i+1} + n_i}{2}. \tag{3.54}$$

In Eq. (3.53), the assumed linearity of the potential between meshes, is implied by the use of the centered finite differences to express the field on the mid-point. We can now rewrite Eq. (3.53) including the approximation in Eq. (3.54) as

$$J_{i+\frac{1}{2}} = n_{i+1} \left[-q \frac{\mu_n}{2} \frac{\Psi_{i+1} - \Psi_i}{\Delta x} + q \frac{D_n}{\Delta x} \right] - n_i \left[\underbrace{q \frac{\mu_n}{2} \frac{\Psi_{i+1} - \Psi_i}{\Delta x}}_{a} + \underbrace{q \frac{D_n}{\Delta x}}_{b} \right]. \tag{3.55}$$

If we assume a condition where $J_n = 0$ (equilibrium) and $a \gg b$ (negligible diffusion), it is easy to see that positive definite nature of the carrier density is not guaranteed, since the solution oscillates as $n_{i+1} \approx -n_i$. Also, it can be shown that for stability we need to have $\Psi_{i+1} - \Psi_i > 2k_B T/q$, which requires very small mesh spacing to be verified.

The approach by Scharfetter and Gummel [52] has provided an optimal solution to this problem, although the mathematical properties of the proposed scheme have been fully recognized much later. We consider again a linear potential variation between neighboring mesh points, which is consistent with the use of finite differences to express the field. We express the current in the interval $[x_i; x_{i+1}]$ as a truncated expansion about the value at the mid-point

$$J_n(x) = J_n\left(x_{i+\frac{1}{2}}\right) + \left(x - x_{i+\frac{1}{2}}\right) \frac{\delta}{\delta x} J_n(x). \tag{3.56}$$

From Eq. (3.56) we obtain a first-order differential equation for J_n which can be solved to provide $n(x)$ in the mesh interval, using as boundary conditions the values of carrier density n_i and n_{i+1}. We obtain

$$n(x) = [1 - g(x, \Psi)]n_i + g(x, \Psi)n_{i+1}; \quad x \in [x_i; x_{i+1}], \tag{3.57}$$

where $g(x, \Psi)$ is the growth function

$$g(x, \Psi) = \left[1 - \exp\left(\frac{\Psi_{i+1} - \Psi_i}{k_B T/q}\frac{x_1 - x_1}{\Delta x}\right)\right] / \left[1 - \exp\left(\frac{\Psi_{i+1} - \Psi_i}{k_B T/q}\right)\right]. \tag{3.58}$$

The result in Eq. (3.57) can be used to evaluate $n(x_{i+1/2})$ for the discretization of the current in Eq. (3.58). It is easy to see that only when $\Psi(i+1) - \Psi(i) = 0$ we have

$$n_{i+1/2} = \left(1 - \frac{1}{2}\right)n_i + \frac{1}{2}n_{i+1} = \frac{n_i + n_{i+1}}{2}. \tag{3.59}$$

The continuity equation can be easily discretized on rectangular uniform and nonuniform meshes using the above results for the currents, because the mesh lines are aligned exactly.

3.1.9 Extension of the Validity of the Drift–Diffusion Model

Due to the relative simplicity of the DD equations, it would be very appealing to extend the validity of DD-like models well into the hot electron regime. We have seen that the simplest attempt to include high-field effects is to make the mobility and the diffusion coefficient field dependent. The electron current in 1D is

$$J(x, t) = q n(x, t)\mu(E)E + q D(E)\frac{\delta n(x, t)}{\delta x}. \tag{3.60}$$

Here $\mu(E)E = v(E)$ is the (steady-state) drift velocity for the case of homogeneous field E. The field is also space and time dependent, i.e., $E = E(x, t)$. Mobility and diffusion coefficients are steady-state quantities, but the carrier velocity may differ considerably from the steady-state value $v(E)$, due to abrupt space or time changes of the electric field. The steady-state $v(E)$ can be considered accurate only if space or time variations of the field are very smooth (adiabatic). Velocity overshoot occurs when the average electron velocity exceeds the steady-state (bulk) velocity. A modified version of the DD equation to include velocity overshoot was proposed by Thornber [53]:

$$J(x, t) = q n(x, t)\left[v(E) + W(E)\frac{\delta E}{\delta x} + B(E)\frac{\delta E}{\delta t}\right] + q D(E)\frac{\delta n(x, t)}{\delta x} + q A(E)\frac{\delta n(x, t)}{\delta t}, \tag{3.61}$$

where three new terms have been added. The term with $W(E)$ contains the field gradient and corrects the local drift velocity for spatial velocity overshoot effects. The term with $B(E)$

contains the time derivative of the field and corrects for time-dependent velocity overshoot. The last term with $A(E)$ preserves the invariances of the total current (note that this term does not represent generation, recombination, trapping, etc., effects which may be incorporated with an additional term). The quantities $W(E)$, $B(E)$, and $A(E)$ must be tabulated from detailed transport calculations. In steady state the current equation simply becomes

$$J(x) = q\,n(x)\left[v(E) + W(E)\frac{\delta E}{\delta x}\right] + q\,D(E)\frac{\delta n}{\delta x} \qquad (3.62)$$

and the resulting continuity equation is

$$\frac{\delta n(x)}{\delta x} = \frac{\delta}{\delta x}\left[n(x)v(E) + n(x)W(E)\frac{\delta E}{\delta x} + D(E)\frac{\delta n}{\delta x}\right]. \qquad (3.63)$$

Equation (3.63) does not describe a real transient, since the time derivatives in the current equation have been neglected. Therefore, Eq. (3.63) is valid in the steady-state limit $\delta n/\delta t = 0$, i.e., $t \to \infty$. The equation may be used, however, to solve the pseudo-time-dependent problem until steady state is achieved. Since the Poisson equation is solved at each time step, the fields and the related variable are continuously updated in space. Alternatively, one may solve the steady-state equation obtained with $\delta n/\delta t = 0$, using Newton's method, for instance. Developing the space derivatives, Eq. (3.63) becomes [54]

$$\frac{\delta n}{\delta t} = \frac{\delta n}{\delta x}v(E) + \frac{\delta v(E)}{\delta x}n(x) + \frac{\delta n}{\delta x}W(E)E_x + n(x)\frac{\delta}{\delta x}\left[W(E)\frac{\delta E}{\delta x}\right] + D(E)\frac{\delta^2 n}{\delta x^2} + \frac{\delta D(E)}{\delta x}\frac{\delta n}{\delta x}$$

$$= \left(\frac{\delta v(E)}{\delta x} + \frac{\delta}{\delta x}\left[W(E)\frac{\delta E}{\delta x}\right]\right)n(x) + \left[v(E) + W(E)\frac{\delta E}{\delta x} + \frac{\delta D(E)}{\delta x}\right]\frac{\delta n}{\delta x} + D(E)\frac{\delta^2 n}{\delta x^2}.$$

$$(3.64)$$

Finally, the pseudo-time-dependent equation has form

$$\frac{\delta n}{\delta t} = a(x,t)\frac{\partial^2 n}{\partial x^2} + b(x,t)\frac{\partial n}{\partial x} + c(x,t)n. \qquad (3.65)$$

Since from Gauss' law $\delta E/\delta x = \rho/\varepsilon$ and ρ depends on the carrier density $n(x)$, the coefficients $b(x,t)$ and $c(x,t)$ are also functions of $n(x)$ and Eq. (3.64) is nonlinear. Often the overshoot parameter $W(E)$ is rewritten in terms of mobility, as $W(E) = \mu(E)L(E)$, where $L(E)$ is called length coefficient. Monte Carlo calculations as well as analytical models for the length coefficient have been presented in the literature. Extension to 2D is not trivial when confining fields (barriers) besides accelerating fields are present. An approximate approach allows the extension to 2D by using the gradient of the quasi-Fermi levels (very flat inside barrier regions but following the potential profile in accelerating regions) as the functional parameter for the length coefficient [55].

PROBLEMS FOR CHAPTER 3

1. Plot the doping dependence of the low-field electron mobility, as described by the conventional mobility model, Klaassen's mobility model, the Arora model and the Dorkel and Leturg model. In your calculation use the parameters specified in the Silvaco ATLAS manual or the corresponding paper. Consider n-type semiconductor with donor doping density varying from 10^{14} cm^{-3} to 10^{20} cm^{-3}.

2. Assume that the doping density is $N_D = 10^{17}$ cm^{-3}. Plot the field-dependent mobility using each of the above-described low-field mobility models and the expressions for the field-dependent mobility and saturation velocity. Vary the electric field value between 0.1 kV cm^{-1} and 100 kV cm^{-1}. Compare your results with those obtained with the Scharfetter and Gummel model, described by expressions (4.1–50) and (4.1–23) in Selberherr's book.

3. Derive the expression for the SRH generation/recombination rate given by Equation (3.39). What are the limiting values to this expression under low and high injection conditions and when and where do they occur?

4. Plot the perpendicular field dependence of the low-field electron mobility using the Yamaguchi and the Shirahata models. In your calculations assume n-channel MOSFET device with uniform substrate doping equal to 3.9×10^{15} cm^{-3}, 2×10^{16} cm^{-3}, 7.2×10^{16} cm^{-3} and 3×10^{17} cm^{-3}. Vary the transverse electric field from 10^4 V cm^{-1} to 10^6 V cm^{-1}. Compare your model results with the experimental data of Takagi, Toriumi, Iwase, and Tango (IEEE Trans. Electron Devices, Vol. 41, pp. 2357–2362, 1994).

5. To obtain diagonaly-dominant coefficient matrix when using finite-difference scheme for the discretization of the Poisson equation, it is necessary to use some linearization scheme. The simplest way to achieve this is to use $\Psi \rightarrow \Psi + \delta$, where δ is small.

    (a) Write down (derive) the linearized Poisson equation using this linearization scheme.

    (b) Write down (derive) the scaled version of the result obtained in (a).

    (c) Write the finite-difference approximation for the scaled Poisson equation. If one solves (c) for the improvement δ, show that the resultant coefficient matrix A is diagonally dominant. (Note: Matrix A is diagonally dominant if the absolute value of the sum of the off-diagonal elements in each row is smaller than the absolute value of the corresponding diagonal term.)

6. Consider a 1D sample, such that for $x < x_b$ the semiconductor has a dielectric constant ε_1, and for $x > x_b$ has dielectric constant ε_2. At the interface between the two semiconductor materials ($x = x_b$), there are no interface charges. Starting from the

condition

$$\varepsilon_1 \left.\frac{\partial \Psi}{\partial x}\right|_{x=x_b} = \varepsilon_2 \left.\frac{\partial \Psi}{\partial x}\right|_{x=x_b}$$

and using Taylor series expansion for Ψ around $x = x_b$ (for $x < x_b$ and, $x > x_b$) calculate the finite-difference approximation of the Poisson equation at $x = x_b$.

7. This example is a demonstration of the fact that explicit numerical integration methods are incapable of solving even the problem of linearly-graded junctions in thermal equilibrium, for which $N_D - N_A = mx$, where a is the edge of the depletion region. To demonstrate this, calculate the following:

- Establish the boundary conditions for the electrostatic potential [$\Psi(-a)$ and $\Psi(a)$] by taking into account the free carrier terms in the equilibrium 1D Poisson equation:

$$\frac{\partial^2 \Psi}{\partial x^2} = -\frac{e}{\varepsilon}(p - n + mx) = -\frac{e}{\varepsilon}(n_i e^{-\Psi/V_t} - n_i e^{\Psi/V} + mx)$$

- Solve analytically the 1D Poisson equation for $\Psi(x)$ within the depletion approximation (no free carriers) and calculate a using this result as well as the boundary conditions found in (step 1). What is the expression for absolute value of the maximum electric field.

- Apply the explicit integration method for the numerical solution of the 1D Poisson equation (that includes the free carriers) by following the steps outlined below:

  ◦ Write a Taylor series expansion for $\Psi(x)$ around $x = 0$, keeping the terms up to the fifth order.

  ◦ Starting from the equilibrium Poisson equation, analytically calculate $\Psi''(0)$, $\Psi^{(3)}(0)$, $\Psi^{(4)}(0)$, and $\Psi^{(5)}(0)$.

  ◦ Use the maximum value of the electric field derived in (step 2) to determine from the Taylor series expansion for $\Psi(x)$, the terms $\Psi(h)$, $\Psi(2h)$, and $\Psi(3h)$.

  ◦ Compute $\Psi(x)$ at $x = 4h, 5h, 6h, \ldots$, up to $x_{max} = 0.5$ μm, using the predictor–corrector method in which the predictor formula:

  $$\Psi_{i+1} = 2\Psi_{i-1} - \Psi_{i-3} + 4h^2 \left(\Psi_{i-1}'' + \frac{\Psi_i'' - 2\Psi_{i-1}'' + \Psi_{i-2}''}{3} \right)$$

  ◦ is applied to predict Ψ_{i+1}, which is then corrected by the corrector formula:

  $$\Psi_{i+1} = 2\Psi_i - \Psi_{i-1} + h^2 \left(\Psi_i'' + \frac{\Psi_{i+1}'' - 2\Psi_i'' + \Psi_{i-1}''}{12} \right)$$

◦ In both, the predictor and the corrector formulas, the second derivatives are obtained from the Poisson's equation. The role of the predictor is to provide Ψ''_{i+1} that appears in the corrector formula.

• Repeat the above procedure for the following values of the first derivative:

◦ Trial 1: $\Psi'(0)_1 = \Psi'(0)$,

◦ Trial 2: $\Psi'(0)_2 = \Psi'(0)_1/2$,

◦ Trial 3: $\Psi'(0)_3 = 0.5[\Psi'(0)_1 + \Psi'(0)_2]$.

◦ Repeat the above described process for several iteration numbers, say up to $n = 22$. Comment on the behavior of this explicit integration scheme.

Use the following parameters in the numerical integration:

$$e = 1.602 \times 10^{-19}\text{C}, \, \varepsilon = 12\varepsilon_0 = 1.064 \times 10^{-12} \text{ F cm}^{-1}, \, T = 300 \text{ K},$$
$$n_i = 1.4 \times 10^{10}\text{cm}^{-3}, \, m = 10^{21}\text{cm}^{-4}, \, h = 2 \times 10^{-7}\text{cm}.$$

8. Write a 1D Poisson equation solver that solves the linearized Poisson equation for a pn-junction under equilibrium condition, with:

(a) $N_A = 10^{15}\text{cm}^{-3}$, $N_D = 10^{15}\text{cm}^{-3}$,

(b) $N_A = 10^{16}\text{cm}^{-3}$, $N_D = 10^{16}\text{cm}^{-3}$,

(c) $N_A = 10^{16}\text{cm}^{-3}$, $N_D = 10^{18}\text{cm}^{-3}$.

For each of these three device structures:

− Calculate analytically the maximum allowed mesh size in each region.
− Plot the conduction band edge versus distance assuming that the Fermi level is the reference energy level ($E_F = 0$).
− Plot the total charge density versus distance.
− Plot the electric field versus distance. Calculate the electric field using centered difference scheme.
− Plot electron and hole densities versus distance.
− Calculate analytically, using the block-charge approximation, the width of the depletion regions and the magnitude of the peak electric field, and compare the analytical with the numerical simulation results. When is the block-charge approximation invalid?

9. Develop a one-dimensional (1D) DD simulator for modeling pn-junctions (diodes) under forward and reverse bias conditions. Include both types of carriers in your model (electrons and holes). Derive the finite-difference expressions for the electron and hole

current continuity equations using Sharfetter–Gummel discretization scheme, which was described in the class. Compare your final answers with the expressions given in Selberherr's book.

Model:

Silicon diode, with permittivity $\varepsilon_{sc} = 1.05 \times 10^{-10}$ F/m and intrinsic carrier concentration $n_i = 1.5 \times 10^{10}$ cm^{-3} at $T = 300$ K. In all your simulations assume that $T = 300$ K. Use concentration-dependent and field-dependent mobility models and SRH generation–recombination process. Assume ohmic contacts and charge neutrality at both ends to get the appropriate boundary conditions for the potential and the electron and hole concentrations. For the concentration-dependent electron and hole mobilities use the Arora model given in the Silvaco Manuals. It will be easier, while you are developing the 1D simulator, to assume constant mobility (concentration and field-independent) model. For the field-dependent mobility, use the model described in the class notes and in the Silvaco ATLAS manual, using the relevant parameters listed there. For the SRH generation–recombination, use the expressions given in the text. To simplify your calculations, assume that the trap energy level coincides with the intrinsic level.

Doping: Use $N_A = 10^{16}$ cm^{-3} and $N_D = 10^{17}$ cm^{-3} as a net doping of the p- and n-regions, respectively.

Poisson solver:

Modify your solver that you developed in Problem #8 to solve the linear Poisson equation:

$$\frac{\partial^2 \psi}{\partial x^2} = -\frac{e}{\varepsilon_{sc}} [p - n + N_D(x) - N_A(x)]$$

Numerical methods: Use the LU decomposition method for the solution of the 1D Poisson and the two 1D continuity equations for electrons and holes individually. Use Gummel's decoupled scheme, described in the class and in the distributed notes, to solve the resultant set of coupled set of algebraic equations.

Outputs:

- Plot the conduction band edge under equilibrium conditions (no current flow).

- Vary the anode bias V_A from 0 to 0.6 V, in voltage increments that are fraction of the thermal voltage $V_T = k_B T/q$, to have stable convergence. Plot the resulting IV characteristics. The current will be in A/unit area, since you are doing 1D modeling. Check the conservation of current when going from the cathode to the anode, which also means conservation of particles in your system. For the calculation of the current density, use the results given in the notes.

- For $V_A = 0.5$ V, plot the position of the electron and hole quasi-Fermi levels, with respect to the equilibrium Fermi level, assumed to be the reverence energy level.

CHAPTER 4

Hydrodynamic Model

The current drive capability of deeply scaled MOSFETs and, in particular, n-MOSFETs has been the subject of investigation since the late 1970s. First it was hypothesized that the effective carrier injection velocity from the source into the channel would reach the limit of the saturation velocity and remain there as longitudinal electric fields increased beyond the onset value for velocity saturation. However, theoretical work indicated that velocity overshoot can occur even in silicon [56], and indeed it is routinely seen in the high-field region near the drain in simulated devices using energy balance models or Monte Carlo simulation. While it was understood that velocity overshoot near the drain would not help current drive, experimental work [57, 58] claimed to observe velocity overshoot near the source, which would be beneficial and render the drift-diffusion (DD) model invalid.

In the computational electronics community, the necessity for the hydrodynamic (HD) transport model is normally checked by comparison of simulation results for HD and DD simulations. Despite the obvious fact that, depending on the equation set, different principal physical effects are taken into account, the influence on the models for the physical parameters is more subtle. The main reason for this is that in the case of the HD model, information about average carrier energy is available in form of the carrier temperature. Many parameters depend on this average carrier energy, e.g., the mobilities and the energy relaxation times. In the case of the DD model, the carrier temperatures are assumed to be in equilibrium with the lattice temperature, that is $T_C = T_L$, hence, all energy dependent parameters have to be modeled in a different way.

4.1 EXTENSIONS OF THE DRIFT-DIFFUSION MODEL

The DD model is the simplest current transport model which can be derived from Boltzmann's transport equation (BTE) by the method of moments [59] as discussed in Chapter 3, or from basic principles of irreversible thermodynamics [60]. For many decades, the DD model has been the backbone of semiconductor device simulation. As was discussed in detail in Chapter 3, the electron current density is phenomenologically expressed as consisting of two components. The drift component is driven by the electric field and the diffusion component by the electron density gradient. It is given by

$$J = q(n\mu_n E + D_n \nabla n),$$ (4.1)

where μ and D_n are the mobility and the diffusivity of the electron gas, respectively, and are related to each other by the Einstein relation for nondegenerate semiconductors

$$D_n = \frac{k_B T_n}{q} \mu n,$$ (4.2)

where k_B is the Boltzmann constant. The current relation Eq. (4.1) is inserted into the continuity equation

$$\nabla \cdot J = q \partial_t n$$ (4.3)

to give a second-order parabolic differential equation, which is then solved together with Poisson's equation. Note that generation/recombination effects were neglected in Eq. (4.3) for simplicity.

In the DD approach, the electron gas is assumed to be in thermal equilibrium with the lattice temperature ($T_n = T_L$). However, in the presence of a strong electric field, electrons gain energy from the field and the temperature T_n of the electron gas is elevated. Since the pressure of the electron gas is proportional to $nk_B T_n$, the driving force now becomes the pressure gradient rather then merely the density gradient. This introduces an additional driving force, namely, the temperature gradient besides the electric field and the density gradient. Phenomenologically, one can write

$$J = q(n\mu_n E + D_n \nabla n + n D_T \nabla T_n),$$ (4.4)

where D_T is the thermal diffusivity.

Although the DD equations are based on the assumption that the electron gas is in thermal equilibrium with the lattice, an estimation for the local temperature can be calculated with the local energy balance equation [61]

$$T_n = T_L + \frac{2}{3} \frac{q}{k_B} \tau_\varepsilon \mu E^2 = T_L \left[1 + \left(\frac{E}{E_c} \right)^2 \right],$$ (4.5)

where τ_ε is the energy relaxation time. Equation (4.5) is obtained under the assumption of a local energy balance. At the critical electric field, E_c, which depends on the electric field via the mobility, the carrier temperature reaches twice the value of the lattice temperature. E_c is on the order of $10\ kV\ cm^{-1}$, a value easily exceeded even in relatively long channel devices where values higher than $1\ MV\ cm^{-1}$ can be observed [62]. Note too that the temperature obtained from Eq. (4.4) introduces an inconsistency with the assumptions made during the derivation of the DD model where the electron gas has been assumed to be in equilibrium with the lattice temperature.

For a rapidly varying electric field, however, the average energy lags behind the electric field, and the assumption of local equilibrium becomes invalid [63]. A consequence of the lag is that the maximum energy can be considerably smaller than the one predicted by the local energy balance equation. An important consequence of this behavior is that the lag of the average energy gives rise to an overshoot in the carrier velocity. The reason for the velocity overshoot is that the mobility depends to first order on the average energy and not on the electric field. As the mobility has not yet been reduced by the increased energy but the electric field is already large, an overshoot in the velocity is observed until the carrier energy comes into equilibrium with the electric field again.

Similar to the carrier mobility, many other physical processes like impact ionization are more accurately described by a local energy model rather than a local electric field model, because these processes depend on the distribution function rather than on the electric field. Altogether, it can be noted that modeling of deep-submicrometer devices with the DD model is becoming more and more problematic. Although successful reproduction of terminal characteristics of nanoscale MOS transistors has been reported with the DD model [64], the values of the physical parameters used significantly violate basic physical principles. In particular, the saturation velocity had to be set to more than twice the value observed in bulk measurements. This implies that the model is no longer capable of reproducing the results of bulk measurements and as such looses its consistency. Furthermore, the model can hardly be used for predictive simulations. These solutions may provide short-term fixes to available models, but obtaining "correct" results from the wrong physics is unsatisfactory in the long run.

In the following, we first give a brief description of the Stratton's approach, which has been one of the first attempts to address all of the above issues. Then, we present the derivation and discuss the properties of the HD model and its simplifications from the moments of the Boltzmann's transport equation introduced in Chapter 2 of this book.

4.2   STRATTON'S APPROACH

One of the first derivations of extended transport equations was performed by Stratton [65]. First the distribution function is split into the even and odd parts

$$f(\mathbf{k}, \mathbf{r}) = f_0(\mathbf{k}, \mathbf{r}) + f_1(\mathbf{k}, \mathbf{r}). \tag{4.6}$$

From the fact that f_1 is odd, $f_1(-\mathbf{k}, \mathbf{r}) = -f_1(\mathbf{k}, \mathbf{r})$, it follows that $\langle f_1 \rangle = 0$. Assuming that the collision operator C is linear and invoking the microscopic relaxation time approximation for the collision operator

$$C[f] = -\frac{f - f_{eq}}{\tau(\varepsilon, \mathbf{r})} \qquad (4.7)$$

the BTE can be split into two coupled equations. In particular, f_1 is related to f_0 via

$$f_1 = -\tau(\varepsilon, \mathbf{r}) \left(\mathbf{v} \cdot \nabla_{\mathbf{r}} f_0 - \frac{q}{\hbar} \mathbf{E} \cdot \nabla_{\mathbf{k}} f_0 \right). \qquad (4.8)$$

The microscopic relaxation time is then expressed by a power law

$$\tau(\varepsilon) = \tau_0 \left(\frac{\varepsilon}{k_B T_L} \right)^{-p}. \qquad (4.9)$$

When f_0 is assumed to be a heated Maxwellian distribution, the following system of equations is obtained

$$\nabla \cdot \mathbf{J} = q \frac{\partial n}{\partial t},$$
$$\mathbf{J} = q n \mu \mathbf{E} + k_B \nabla (n \mu T_n),$$
$$\nabla \cdot (n\mathbf{S}) = -\frac{3}{2} k_B \partial (n T_n) + \mathbf{E} \cdot \mathbf{J} - \frac{3}{2} k_B n \frac{T_n - T_L}{\tau_\varepsilon},$$
$$n\mathbf{S} = -\left(\frac{5}{2} - p \right) \left(\mu n k_B T_n \mathbf{E} + \frac{k_B^2}{q} \nabla (n \mu T_n) \right). \qquad (4.10)$$

Equation (4.4) for the current density can be rewritten as

$$J = q\mu \left(nE + \frac{k_B}{q} T_n \nabla n + \frac{k_B}{q} n (1 + v_n) \nabla T_n \right) \qquad (4.11)$$

with

$$v_n = \frac{T_n}{\mu} \frac{\partial \mu}{\partial T_n} = \frac{\partial \ln \mu}{\partial \ln T_n}, \qquad (4.12)$$

which is commonly used as a fit parameter with values in the range $[-0.5, -1.0]$. For $v_n = -1.0$, the thermal distribution term disappears. The problem with Eq. (4.9) for τ is that p must be approximated by an average value to cover the relevant processes. In the particular case of impurity scattering, p can be in the range $[-1.5, 0.5]$, depending on charge screening. Therefore, this average depends on the doping profile and the applied field; thus, no unique value for p can be given. Note also that the temperature T_n is a parameter of the heated Maxwellian distribution, which has been assumed in the derivation. Only for parabolic bands and a Maxwellian distribution, is this parameter equivalent to the normalized second-order moment.

4.3 BALANCE EQUATIONS MODEL

In this section, we will first describe the prescription for generating the balance equations from the BTE (Eq. 2.31), which can be rewritten in the form

$$\frac{\partial f}{\partial t} + \boldsymbol{\nabla}_r \cdot (vf) - \frac{e}{\hbar}\mathbf{E} \cdot \boldsymbol{\nabla}_k f = S_{\text{op}}f + s(\mathbf{r}, \mathbf{p}, t). \qquad (4.13)$$

We define a quantity $\phi(\mathbf{p})$, that can have values 1, $\mathbf{p}$, ..., etc., and the total value (averaged) of the quantity associated with $\phi(\mathbf{p})$ is

$$n_\phi(\mathbf{r}, t) = \frac{1}{V}\sum_{\mathbf{p}}\phi(\mathbf{p})f(\mathbf{r}, \mathbf{p}, t), \qquad (4.14)$$

where n_ϕ can represent carrier density, current density, etc. To find the balance equation for n_ϕ, we need to multiply the BTE by $\phi(\mathbf{p})/V$ and to integrate over $\mathbf{p}$. Then, the various terms that appear in the BTE become

$$\frac{1}{V}\sum_{\mathbf{p}}\phi(\mathbf{p})\frac{\partial f}{\partial t} = \frac{\partial}{\partial t}\left[\frac{1}{V}\sum_{\mathbf{p}}\phi(\mathbf{p})f(\mathbf{r}, \mathbf{p}, t)\right] = \frac{\partial n_\phi}{\partial t},$$

$$\frac{1}{V}\sum_{\mathbf{p}}\phi(\mathbf{p})\boldsymbol{\nabla}_{\mathbf{r}} \cdot (\mathbf{v}f) = \boldsymbol{\nabla}_{\mathbf{r}} \cdot \left[\frac{1}{V}\sum_{\mathbf{p}}\mathbf{v}\phi(\mathbf{p})f(\mathbf{r}, \mathbf{p}, t)\right] = \boldsymbol{\nabla}_{\mathbf{r}} \cdot F_\phi, \qquad (4.15)$$

where $F_\phi = \frac{1}{V}\sum_{\mathbf{p}}\mathbf{v}\phi(\mathbf{p})f(\mathbf{r}, \mathbf{p}, t)$ is the flux associated with n_ϕ. For example, if $\phi(\mathbf{p}) = 1$, then F_ϕ is the carrier flux, and if $\phi(\mathbf{p}) = E_p$, then F_ϕ is the energy flux. The third term is then given by

$$-e\sum_{\mathbf{p}}\phi(\mathbf{p})\mathbf{E} \cdot \boldsymbol{\nabla}_{\mathbf{p}}f = -e\mathbf{E} \cdot \sum_{\mathbf{p}}\phi(\mathbf{p})\boldsymbol{\nabla}_{\mathbf{p}}f$$

$$= e\mathbf{E} \cdot \sum_{\mathbf{p}}f(\mathbf{r}, \mathbf{p}, t)\boldsymbol{\nabla}_{\mathbf{p}}\phi(\mathbf{p}) = -G_\phi \qquad (4.16)$$

and is called a generation term since electric field increases momentum and the quantity n_ϕ. There is another term $s(\mathbf{r}, \mathbf{p}, t)$ that leads to increase/decrease of n_ϕ (generation–recombination process). The contribution of these terms is

$$S_\phi(\mathbf{r}, t) = \frac{1}{V}\sum_{\mathbf{p}}\phi(\mathbf{p})s(\mathbf{r}, \mathbf{p}, t). \qquad (4.17)$$

The collision events, for example, destroy momentum and thus represent a recombination term

$$R_\phi = -\frac{1}{V}\sum_{\mathbf{p}}\phi(\mathbf{p})\left.\frac{\partial f}{\partial t}\right|_{\text{coll}} = \left\langle\!\!\left\langle\frac{1}{\tau_\phi}\right\rangle\!\!\right\rangle\left[n_\phi(\mathbf{r}, t) - n_\phi^0(\mathbf{r}, t)\right], \qquad (4.18)$$

where $\langle\langle 1/\tau_\phi \rangle\rangle$ is the ensemble relaxation rate. To find the definition of this term we assume nondegenerate semiconductor, for which

$$\frac{\partial f}{\partial t}\bigg|_{\text{coll}} = \sum_{p'} \left[f(\mathbf{r}, \mathbf{p'}, t)S(\mathbf{p'}, \mathbf{p}) - f(\mathbf{r}, \mathbf{p}, t)s(\mathbf{p}, \mathbf{p'}) \right]. \qquad (4.19)$$

Then:

$$\frac{1}{V} \sum_{\mathbf{p}} \phi(\mathbf{p}) \frac{\partial f}{\partial t}\bigg|_{\text{coll}} = -\frac{1}{V} \sum_{\mathbf{p}} \phi(\mathbf{p}) f(\mathbf{r}, \mathbf{p}, t) \sum_{\mathbf{p'}} \left[1 - \frac{\phi(\mathbf{p'})}{\phi(\mathbf{p})} \right] S(\mathbf{p}, \mathbf{p'})$$
$$= -\frac{1}{V} \sum_{\mathbf{p}} \phi(\mathbf{p}) f(\mathbf{r}, \mathbf{p}, t) \frac{1}{\tau_\phi(\mathbf{p})}, \qquad (4.20)$$

where $1/\tau_\phi(\mathbf{p})$ is the total out-scattering rate associated with quantity ϕ. With some manipulation of the above expressions, we get

$$\left\langle\!\left\langle \frac{1}{\tau_\phi} \right\rangle\!\right\rangle = \frac{\dfrac{1}{V} \displaystyle\sum_{\mathbf{p}} \dfrac{f(\mathbf{r}, \mathbf{p}, t)\phi(\mathbf{p})}{\tau_\phi(\mathbf{p})}}{n_\phi(\mathbf{r}, t) - n_\phi^0(\mathbf{r}, t)}. \qquad (4.21)$$

This ensemble relaxation rate depends upon the type of the scattering mechanism and how carriers are distributed in momentum. In summary, when $\phi(\mathbf{p}) = p_i$, then $n_\phi = P_i$ and

$$\frac{dP_i}{dt}\bigg|_{\text{coll}} = -P_i \left\langle\!\left\langle \frac{1}{\tau_m} \right\rangle\!\right\rangle. \qquad (4.22)$$

Also, when $\phi(\mathbf{p}) = E(\mathbf{p})$, then $n_\phi = W$ (average kinetic energy density) which is given by $W = nu$ (u equals the average energy density per electron). Then

$$\frac{dW}{dt}\bigg|_{\text{coll}} = -\left\langle\!\left\langle \frac{1}{\tau_E} \right\rangle\!\right\rangle (W - W_0). \qquad (4.23)$$

Note that the results presented up to this point are exact, i.e. no relaxation time approximation as discussed in Chapter 3 is made. By summarizing all of the above results, we arrive at the balance equation for the quantity n_ϕ, in the form

$$\frac{\partial n_\phi}{\partial t} = -\nabla \cdot F_\phi + G_\phi - R_\phi + S_\phi. \qquad (4.24)$$

a) Carrier-density balance equation

When $\phi(\mathbf{p}) = 1$, then $n_\phi = n$ (electron density), $F_\phi = \frac{1}{V}\sum_{\mathbf{p}} \mathbf{v}f(\mathbf{r}, \mathbf{p}, t) = -\frac{1}{e}\mathbf{J}_n(\mathbf{r}, t)$, $G_\phi = R_\phi = 0$ (scattering mechanisms and the electric field redistribute carriers among states but do not generate or destroy carriers). The above results lead to the continuity equation for the electrons, which is nothing more than conservation of particles in the system

$$\frac{\partial n}{\partial t} = \frac{1}{e} \boldsymbol{\nabla} \cdot \mathbf{J}_n + S_n. \tag{4.25}$$

b) The momentum balance equation

The momentum balance equation is obtained by assuming $\phi(\mathbf{p}) = p_z$, for example. Then, the various quantities that appear in the balance equation are of the following form

$$n_\phi = \frac{1}{V} \sum_\mathbf{p} p_z f(\mathbf{r}, \mathbf{p}, t) = P_z,$$

$$\mathbf{F}_\phi = \frac{1}{V} \sum_\mathbf{p} \mathbf{v} p_z f(\mathbf{r}, \mathbf{p}, t) \rightarrow F_{\phi i} = \frac{1}{V} \sum_\mathbf{p} m^* v_z v_i f. \tag{4.26}$$

In the above expressions P_z is the total momentum density along z, and $F_{\phi i} = 2W_{iz}$, where W_{iz} is a component of the kinetic energy density tensor. The generation and the recombination term reduce to

$$G_\phi = -e\mathbf{E} \cdot \sum_\mathbf{p} f(\mathbf{r}, \mathbf{p}, t) \nabla_\mathbf{p} p_z = -e\, E_z n \,, \tag{4.27}$$

$$R_\phi = P_z \left\langle\!\!\left\langle \frac{1}{\tau_m} \right\rangle\!\!\right\rangle. \tag{4.28}$$

Then, the momentum balance equation for P_z reads

$$\frac{\partial P_z}{\partial t} = -\boldsymbol{\nabla} \cdot F_\phi - enE_z - \left\langle\!\!\left\langle \frac{1}{\tau_m} \right\rangle\!\!\right\rangle P_z = -\sum_i 2\frac{\partial W_{iz}}{\partial x_i} - enE_z - \left\langle\!\!\left\langle \frac{1}{\tau_m} \right\rangle\!\!\right\rangle P \tag{4.29}$$

The trace of the tensor for the total energy density $\overset{\leftrightarrow}{W}$ is

$$\mathrm{Tr}\left(\overset{\leftrightarrow}{W}\right) = \sum_i W_{ii} = \sum_i \frac{1}{2V} \sum_\mathbf{p} m^* v_i^2 f = \frac{1}{V} \sum_\mathbf{p} f(r, p, t) \sum_i \frac{1}{2} m^* v_i^2 = W = nu. \tag{4.30}$$

For simple parabolic bands, we have $\mathbf{P} = nm^*\mathbf{v}_\mathrm{d} = -\mathbf{J}m^*/e$, which gives $\mathbf{J} = -e\mathbf{P}/m^*$ or

$$\frac{\partial J_z}{\partial t} = \frac{2e}{m^*} \sum_i \frac{\partial W_{iz}}{\partial x_i} + \frac{ne^2}{m^*} E_z - \left\langle\!\!\left\langle \frac{1}{\tau_m} \right\rangle\!\!\right\rangle J_z. \tag{4.31}$$

With appropriate simplifications, the balance equation for the current density reduces to the DD equation as discussed later in this section.

c) Energy balance equation

The energy balance equation is obtained from the prescription that $\phi(\mathbf{p}) = E_p$. Then, the expression for the total energy density is found from

$$n_\phi = \frac{1}{V} \sum_{\mathbf{p}} E(\mathbf{p}) f(\mathbf{r}, \mathbf{p}, t) = W \qquad (4.32)$$

and the energy flux is given by

$$F_\phi = \frac{1}{V} \sum_{\mathbf{p}} \mathbf{v} E(\mathbf{p}) f(\mathbf{r}, \mathbf{p}, t) = \mathbf{F}_W. \qquad (4.33)$$

The generation and recombination terms that appear in the balance equation are

$$G_\phi = \mathbf{E} \cdot \left(-\frac{e}{V} \right) \sum_p \mathbf{v} f(r, p, t) = \mathbf{E} \cdot \mathbf{J}_n, \qquad (4.34)$$

$$R_\phi = \left\langle\!\left\langle \frac{1}{\tau_E} \right\rangle\!\right\rangle (W - W_0), \qquad (4.35)$$

where the generation term describes the energy increase due to the electric field and the recombination term gives the loss of energy due to phonons. The final form of the energy balance equation is

$$\frac{\partial W}{\partial t} = -\nabla \cdot \mathbf{F}_W + \mathbf{E} \cdot \mathbf{J} - \left\langle\!\left\langle \frac{1}{\tau_E} \right\rangle\!\right\rangle (W - W_0), \qquad (4.36)$$

which is nothing more than a statement of conservation of energy in the system. The term on the LHS, which describes an increase in energy, is balanced by the terms on the RHS. The first term on the RHS describes the energy flowing into the volume, the second one gives the energy increase due to field accelerating the carriers and the last term describes the energy loss due to collisions.

(d) Complete hydrodynamic equations

Summarizing the previous discussion, the first three balance equations take the form

$$\frac{\partial n}{\partial t} = \frac{1}{e} \nabla \cdot \mathbf{J}_n + S_n,$$

$$\frac{\partial J_z}{\partial t} = \frac{2e}{m^*} \sum_i \frac{\partial W_{iz}}{\partial x_i} + \frac{ne^2}{m^*} E_z - \left\langle\!\left\langle \frac{1}{\tau_m} \right\rangle\!\right\rangle J_z, \qquad (4.37)$$

$$\frac{\partial W}{\partial t} = -\nabla \cdot \mathbf{F}_W + \mathbf{E} \cdot \mathbf{J} - \left\langle\!\left\langle \frac{1}{\tau_E} \right\rangle\!\right\rangle (W - W_0).$$

The balance equation for the carrier density introduces the carrier current density, which in turn introduces the kinetic energy density. The balance equation for the kinetic energy density, on the other hand, introduces the energy flux. Therefore, a new variable appears in the hierarchy of balance equations and the infinite set of balance equations generated in this fashion is actually the solution of the BTE.

The balance equations can be reformulated in a more convenient way by separating the carrier temperature T_C from the lattice temperature T_L. To achieve this, we consider the kinetic energy density tensor and write the carrier velocity as $\mathbf{v} = \mathbf{v}_d + \mathbf{c}$, where the first term describes the average drift velocity and the second term describes the random thermal component. Then, $\langle v_i v_z \rangle = \langle (v_{di} + c_i)(v_{dz} + c_z) \rangle = \langle v_{di} v_{dz} + c_i v_{dz} + c_z v_{di} + c_i c_z \rangle$, where the brackets $\langle \rangle$ represent an average over the distribution function. Now, since $\langle c_i \rangle = 0$, we have $\langle v_i v_z \rangle = \langle v_{di} v_{dz} \rangle + \langle c_i c_z \rangle$. The kinetic energy tensor component W_{iz} is, thus, given by

$$W_{iz} = \frac{1}{2}nm^* \langle v_i v_z \rangle = \frac{1}{2}nm^* \langle v_{di} v_{dz} \rangle + \frac{1}{2}nm^* \langle c_i c_z \rangle, \qquad (4.38)$$

where the first term on the right represents the drift energy K_{iz} and the second term describes the thermal energy due to the random thermal motion of the carriers. The kinetic energy density equals the trace of the tensor $\overset{\leftrightarrow}{W}$, i.e.,

$$W = \sum_{i=x,y,z} W_{ii} = \frac{1}{2}nm^* \sum_i \langle v_{di}^2 \rangle + \frac{1}{2}nm^* \sum_i \langle c_i^2 \rangle,$$
$$= \frac{1}{2}nm^* \langle v_d^2 \rangle + \frac{1}{2}nm^* \langle c^2 \rangle. \qquad (4.39)$$

For the thermal carrier energy we have

$$\frac{1}{2}nm^* \langle c^2 \rangle = \frac{1}{2}nm^* \sum_i \langle c_i^2 \rangle = \frac{3}{2}nk_B T_C = \frac{3}{2}nm^* \langle c_i^2 \rangle. \qquad (4.40)$$

Therefore

$$\frac{1}{2}nm^* \langle c_i^2 \rangle = \frac{1}{2}nk_B T_C, \langle c_i c_j \rangle = \frac{k_B}{m^*} T_{ij} \qquad (4.41)$$

and

$$W_{iz} = \frac{1}{2}nm^* \langle v_{di} v_{dz} \rangle + \frac{1}{2}nm^* \langle c_i c_z \rangle = \frac{1}{2}nm^* \langle v_{di} v_{dz} \rangle + \frac{1}{2}nm^* T_{iz} \frac{k_B}{m^*}, \qquad (4.42)$$

where T_{iz} is a component of the temperature tensor.

We now want to express the energy flux in terms of the temperature tensor. The energy flux, defined earlier and repeated here for convenience, is calculated using

$$\mathbf{F}_W = \frac{1}{V} \sum_{\mathbf{p}} \mathbf{v} E(\mathbf{p}) f(\mathbf{r}, \mathbf{p}, t), \qquad (4.43)$$

which means that the ith component of this vector equals

$$F_{Wi} = \frac{1}{V} \sum_{\mathbf{p}} v_i \frac{m^* v^2}{2} f(\mathbf{r}, \mathbf{p}, t) = \frac{nm^*}{2} \langle v_i v^2 \rangle,$$

$$= \frac{nm^*}{2} \left[v_{di} \sum_j \left(\langle v_{dj}^2 \rangle + \langle c_j^2 \rangle \right) + 2 \sum_j v_{dj} \langle c_i c_j \rangle + \sum_j \langle c_i c_j^2 \rangle \right], \qquad (4.44)$$

$$= v_{di} W + nk_B \sum_j T_{ij} v_{dj} + Q_i,$$

where Q_i is the component of the heat flux vector, which describes the loss of energy due to heat flow out of the volume. To summarize, the kinetic energy flux equals the sum of the kinetic energy density times velocity plus the velocity times the pressure, which actually represents the work to push the volume plus the loss of energy due to flow of heat out. In mathematical terms this is expressed as

$$\mathbf{F}_W = \mathbf{v} W + nk_B \overleftrightarrow{T} \cdot \mathbf{v} + \mathbf{Q}. \qquad (4.45)$$

With the above considerations, the momentum and the energy balance equations reduce to

$$\frac{\partial J_z}{\partial t} = \frac{2e}{m^*} \sum_i \frac{\partial}{\partial x_i} \left(K_{iz} + \frac{1}{2} nk_B T_{iz} \right) + \frac{ne^2}{m^*} E_z - \left\langle\!\left\langle \frac{1}{\tau_m} \right\rangle\!\right\rangle J_z$$

$$\frac{\partial W}{\partial t} = -\nabla \cdot \left(\mathbf{v} W + \mathbf{Q} + nk_B \overleftrightarrow{T} \cdot \mathbf{v} \right) + \mathbf{E} \cdot \mathbf{J}_n - \left\langle\!\left\langle \frac{1}{\tau_E} \right\rangle\!\right\rangle (W - W_0). \qquad (4.46)$$

4.3.1 Displaced Maxwellian Approximation

The most common way to solve the balance equations is to guess a form for the distribution function and use the balance equations to solve for the parameters in this functional form. The most commonly used form is the displaced-Maxwellian

$$f(\mathbf{p}) \propto \exp\left[-|\mathbf{p} - m^* \mathbf{v}_d|^2 / 2m^* k_B T_C \right]. \qquad (4.47)$$

This distribution is a good model for cases when electron–electron interactions are strong enough to thermalize the distribution function. For a displaced Maxwellian, the temperature tensor is diagonal, i.e., $T_{ij} = T_C \delta_{ij}$ so that $\sum_i \frac{\partial}{\partial x_i} \left(\frac{1}{2} nk_B T_C \delta_{iz} \right) = \frac{1}{2} \frac{\partial}{\partial x_z} \left(\frac{1}{2} nk_B T_C \right)$. With these simplifications, the current density (momentum) balance equation becomes

$$\frac{\partial J_z}{\partial t} = \frac{2e}{m^*} \sum_i \frac{\partial K_{iz}}{\partial x_i} + \frac{e}{m^*} \frac{\partial}{\partial x_z} (nk_B T_C) + \frac{ne^2}{m^*} E_z - \left\langle\!\left\langle \frac{1}{\tau_m} \right\rangle\!\right\rangle J_z. \qquad (4.48)$$

The momentum balance equation can be immediately obtained by multiplying the above result by $(-e/m^*)^{-1}$ to get

$$\frac{\partial P_z}{\partial t} = -\sum_i 2\frac{\partial K_{iz}}{\partial x_i} - \frac{\partial}{\partial x_z}(nk_B T_C) - enE_z - \left\langle\left\langle\frac{1}{\tau_m}\right\rangle\right\rangle P_z. \qquad (4.49)$$

Therefore, to solve the balance equation for P_z (or equivalently J_z), one needs to know the carrier temperature. We can consider two limiting cases: (1) Under low-field conditions, the carrier temperature T_C can be assumed to be equal to the lattice temperature T_L. (2) Under high-field conditions, the carrier temperature T_C is larger than the lattice temperature, and under these circumstances one needs to solve the energy balance equation that is discussed next.

For the displaced-Maxwellian approximation for the distribution function, the heat flux $\mathbf{Q} = 0$. However, Blotekjaer [66] has pointed out that this term must be significant for non-Maxwellian distributions, so that a phenomenological description for the heat flux, of the form described by the Franz–Wiedermann law is used, which states that

$$\mathbf{Q} = -\kappa\nabla T_C, \qquad (4.50)$$

where κ is the thermal or heat conductivity. In silicon, the experimental value of κ is 14 W (cm K)$^{-1}$. The above description for Q actually leads to a closed set of equations in which the energy balance equation is of the form

$$\frac{\partial W}{\partial t} = -\nabla\cdot(\mathbf{v}W - \kappa\nabla T_C + nk_B T_C\mathbf{v}) + \mathbf{E}\cdot\mathbf{J}_n - \left\langle\left\langle\frac{1}{\tau_E}\right\rangle\right\rangle(W - W_0). \qquad (4.51)$$

It has been recognized in recent years that this approach is not correct for semiconductors in the junction regions, where high and unphysical velocity peaks are established by the Franz–Wiedemann law. To avoid this problem, Stettler, Alam, and Lundstrom [67] have suggested a new form of closure

$$\mathbf{Q} = -\kappa\nabla T_C + \frac{5}{2}(1 - r)\frac{k_B T_L}{e}\mathbf{J}, \qquad (4.52)$$

where $\mathbf{J}$ is the current density and r is a tunable parameter less than unity. Now using

$$\frac{\partial}{\partial x}(2K_{iz}) = \frac{\partial}{\partial x_i}(nm^* v_{di}v_{dz}) = nm^*\frac{\partial}{\partial x}(v_{di}v_{dz})$$

$$= nm^*\left[\frac{\partial v_{di}}{\partial x_i}v_{dz} + v_{dz}\frac{\partial v_{dz}}{\partial x_z}\right] \qquad (4.53)$$

and assuming that the spatial variations are confined along the z-direction, we have

$$\frac{\partial}{\partial x_z}(2K_{iz}) = \frac{\partial}{\partial x_z}\left(nm^* v_{dz}^2\right). \qquad (4.54)$$

Summarizing, the balance equations for the drifted-Maxwellian distribution function simplify to

$$\frac{\partial n}{\partial t} = \frac{1}{e} \nabla \cdot J_n + S_n$$

$$\frac{\partial J_z}{\partial t} = \frac{e}{m^*} \frac{\partial}{\partial x_z} \left(nm^* v_{dz}^2 + nk_B T_C \right) + \frac{ne^2}{m^*} E_z - \left\langle\!\left\langle \frac{1}{\tau_m} \right\rangle\!\right\rangle J_z \qquad (4.55)$$

$$\frac{\partial W}{\partial t} = -\frac{\partial}{\partial x_z} \left[(W + nk_B T_C)\, v_{dz} - \kappa \frac{\partial T_C}{\partial x_z} \right] + J_z E_z - \left\langle\!\left\langle \frac{1}{\tau_E} \right\rangle\!\right\rangle (W - W_0),$$

where

$$J_z = -env_{dz} = -\frac{e}{m^*} P_z$$

$$W = \frac{1}{2} nm^* v_{dz}^2 + \frac{3}{2} nk_B T_C \qquad (4.56)$$

4.3.2   Momentum and Energy Relaxation Rates

Having arrived at the final form of the HD equations, the next task is to calculate the momentum and energy relaxation rates, which in this case are ensemble averaged quantities. For this purpose, one can utilize the drifted-Maxwellian form of the distribution function for simple scattering mechanisms, but for cases where several scattering mechanisms are important, one must use bulk Monte Carlo simulations to calculate these quantities. We consider both cases below.

(A) Drifted-Maxwellian

Assume that the distribution function is of the form given by Equation (4.47)

$$f(\mathbf{p}) = \exp\left[-\frac{|\mathbf{p} - m^* \mathbf{v}_d|^2}{2m^* k_B T_C} \right] \qquad (4.57)$$

where T_C is the carrier temperature and $\mathbf{v}_d = v_{dz}\mathbf{i}_z$. Expanding the distribution function gives

$$f(\mathbf{p}) = \exp\left[-\frac{p^2}{2m^* k_B T_C} \right]\left[1 + \frac{p_z v_{dz}}{k_B T_C} \right] = f_S + f_A. \qquad (4.58)$$

The ensemble averaged momentum relaxation time is then given by

$$\left\langle\!\left\langle \frac{1}{\tau_m} \right\rangle\!\right\rangle = \frac{\sum_\mathbf{p} f(\mathbf{r}, \mathbf{p}, t)\, p_z / \tau_m(\mathbf{p})}{\sum_\mathbf{p} p_z f(\mathbf{r}, \mathbf{p}, t)}. \qquad (4.59)$$

For homogeneous systems $f(\mathbf{r}, \mathbf{p}, t) = f(\mathbf{p})$ (steady-state). Since $\tau_m(\mathbf{p})$ is generally an even function of p_z, we therefore only have contribution from the asymmetric term, which gives

$$\left\langle\!\!\left\langle \frac{1}{\tau_m} \right\rangle\!\!\right\rangle = \frac{\sum_{\mathbf{p}} f_S(E) p_z^2 / \tau_m(\mathbf{p})}{\sum_{\mathbf{p}} p_z^2 f_S(E)}. \qquad (4.60)$$

Now, if θ is the angle between $\mathbf{p}$ and the electric field $\mathbf{E} = E_z \mathbf{i}_z$, we can write $p_z = p \cos \theta$. Also, if we assume a parabolic band structure, for which $E(\mathbf{p}) = p^2 / 2m^*$, then

$$\left\langle\!\!\left\langle \frac{1}{\tau_m} \right\rangle\!\!\right\rangle = \frac{\int E^{3/2} \dfrac{1}{\tau_m(e)} f_S(E) dE}{\int E^{3/2} f_S(E) dE}. \qquad (4.61)$$

For the case that the energy-dependent momentum relaxation rate is of the form $\tau_m(E) = \tau_0 (E/k_B T_L)^S$, we have

$$\left\langle\!\!\left\langle \frac{1}{\tau_m} \right\rangle\!\!\right\rangle = \frac{1}{\tau_0} \left(\frac{T_L}{T_C} \right)^s \frac{\Gamma(5/2 - s)}{\Gamma(5/2)}. \qquad (4.62)$$

For low fields, the standard momentum relaxation rate that enters into the expression for the mobility is given by

$$\langle \tau_m \rangle = \tau_0 \frac{\Gamma(s + 5/2)}{\Gamma(5/2)}, \qquad (4.63)$$

which shows very different behavior from the result given in Eq. (4.62).

Let us consider acoustic phonon scattering for which $s = -1/2$, which then gives $\langle\langle 1/\tau_m \rangle\rangle = A\sqrt{T_C/T_L}$. Similarly, for acoustic phonon scattering the ensemble averaged energy relaxation rate is given by $\langle\langle 1/\tau_E \rangle\rangle = B/T_L\sqrt{T_L/T_C}$. Now, under steady-state conditions and for homogeneous systems, the momentum and energy balance equations become

$$\frac{ne^2}{m^*} E_z = \left\langle\!\!\left\langle \frac{1}{\tau_m} \right\rangle\!\!\right\rangle J_z$$

$$J_z E_z = \left\langle\!\!\left\langle \frac{1}{\tau_E} \right\rangle\!\!\right\rangle (W - W_0), \qquad (4.64)$$

which then leads to

$$T_C = T_L + \frac{2e^2}{3m^* k_B} \frac{E_z^2}{\langle\langle 1/\tau_m \rangle\rangle \langle\langle 1/\tau_E \rangle\rangle} \rightarrow \frac{T_C}{T_L} = 1 + \left(\frac{E_z}{E_{crit}} \right)^2 \qquad (4.65)$$

where E_{crit} is some critical electric field.

(B) Bulk Monte Carlo simulations

An alternative way of deriving the momentum relaxation rate (ensemble averaged) is to use steady-state Monte Carlo simulation for bulk materials under uniform electric fields, which is the topic of Chapter 6. Under these conditions, the momentum and energy balance equations simplify, and where we have that

$$\left\langle\!\left\langle \frac{1}{\tau_m} \right\rangle\!\right\rangle = -\frac{eE_z}{m^* v_{dz}}, \quad \left\langle\!\left\langle \frac{1}{\tau_E} \right\rangle\!\right\rangle = -\frac{e n v_{dz} E_z}{W - W_0}. \qquad (4.66)$$

Note that the as-calculated momentum and energy relaxation rates are electric field dependent, i.e., energy-dependent quantities.

4.3.3   Simplifications that Lead to the Drift-Diffusion Model

To arrive at the DD model, we first rewrite the momentum balance equation in the following form

$$J_z + \frac{1}{\langle\langle 1/\tau_m \rangle\rangle} \frac{\partial J_z}{\partial t} = \frac{e/m^*}{\langle\langle 1/\tau_m \rangle\rangle} \frac{\partial}{\partial x_z} \left(nm^* v_{dz}^2 + nk_B T_C \right) + ne \frac{e/m^*}{\langle\langle 1/\tau_m \rangle\rangle} E_z. \qquad (4.67)$$

Defining the carrier mobility as

$$\mu_n = \frac{e/m^*}{\langle\langle 1/\tau_m \rangle\rangle}, \qquad (4.68)$$

we first have for acoustic deformation potential scattering:

$$\mu_n = \frac{e/m^*}{A\sqrt{T_C/T_L}} = \frac{\mu_0}{\sqrt{1 + (E_z/E_{crit})^2}}, \qquad (4.69)$$

which clearly shows that at high fields, the mobility decreases with increasing the in-plane electric field. We now go back to the momentum balance equation, which we rewrite as

$$J_z + \frac{1}{\langle\langle 1/\tau_m \rangle\rangle} \frac{\partial J_z}{\partial t} = \mu_n \frac{\partial}{\partial x_z} \left(nm^* v_{dz}^2 + nk_B T_C \right) + ne\mu_n E_z. \qquad (4.70)$$

The first approximation that we make is to assume that the carrier drift energy is much smaller than the thermal energy. This approximation is valid for low-field conditions and leads to kinetic energy density of the form

$$W = \frac{1}{2} nm^* v_{dz}^2 + \frac{3}{2} nk_B T_C \approx \frac{3}{2} nk_B T \rightarrow nk_B T_C = \frac{2}{3} W. \qquad (4.71)$$

Under steady-state conditions, the momentum balance equation simplifies to

$$J_z = \mu_n \frac{\partial}{\partial x_z} \left(\frac{2}{3} W \right) + ne\mu_n E_z = ne\mu_n E_z + \frac{2}{3} \mu_n \frac{\partial W}{\partial x_z}. \qquad (4.72)$$

The above expression suggests that diffusion is associated with gradients in the kinetic energy density. The simplified expression for the current density J_z can also be written as

$$J_z = ne\mu_n E_z + \mu_n k_B T_C \frac{\partial n}{\partial x_z} + n k_B \mu_n \frac{\partial T_C}{\partial x_z}$$
$$= ne\mu_n E_z + eD_n + eS_n \frac{\partial T_C}{\partial x_z}, \quad (4.73)$$

where D_n and S_n are the diffusion and the Soret coefficients, respectively. As a further simplification to the DD equations, we assume that there are no temperature gradients in the system. Then, the set of equations that one solves using, for example, the Silvaco simulation software is

$$\frac{\partial n}{\partial t} = \frac{1}{e}\nabla \cdot \mathbf{J}_n + S_n \quad (4.74)$$
$$\mathbf{J}_n = en\mu_n \mathbf{E} + eD_n\nabla n$$

Note that in the above expressions, the diffusion coefficient and the mobility of the carriers are low-field quantities. To extend the validity of this model for high-field conditions, one usually employs field-dependent models for the diffusion coefficient and the mobility. A variety of models have been developed for this purpose and they are summarized in Appendix B.

4.4   NUMERICAL SOLUTION SCHEMES FOR THE HYDRODYNAMIC EQUATIONS

A large class of initial value (time-evolution) PDEs in one space dimension can be cast into the form of a *flux-conservative equation*,

$$\frac{\partial \mathbf{u}}{\partial t} = -\frac{\partial \mathbf{F}(\mathbf{u})}{\partial x}, \quad (4.75)$$

where $\mathbf{u}$ and $\mathbf{F}$ are vectors, and where (in some cases) $\mathbf{F}$ may depend not only on $\mathbf{u}$ but also on spatial derivatives of $\mathbf{u}$. The vector $\mathbf{F}$ is called the *conserved flux*. We will consider, in this section, a prototypical example of the general flux conservative equation above, namely the equation for a scalar u,

$$\frac{\partial u}{\partial t} = -v\frac{\partial u}{\partial x} \quad (4.76)$$

with v a constant. As it happens, we already know analytically that the general solution of this equation is a wave propagating in the positive x-direction

$$u = f(x - vt), \quad (4.77)$$

where f is an arbitrary function. However, the numerical strategies that we develop will be equally applicable to the more general equations represented by Eq. (4.75). In some contexts,

Eq. (4.76) is called an *advective* equation, because the quantity u is transported by a "fluid flow" with a velocity v. How do we go about finite differencing Eq. (4.76)? The straightforward approach is to choose equally spaced points along both the t- and x-axes. Thus denote

$$x_j = x_0 + j\Delta x, \quad j = 0, 1, \dots, J$$
$$t_n = t_0 + n\Delta t \qquad n = 0, 1, \dots, N. \tag{4.78}$$

Let u_j^n denote $u\left(t_n, x_j\right)$. We have several choices for representing the time derivative term. The obvious way is to set

$$\left.\frac{\partial u}{\partial t}\right|_{j,n} = \frac{u_j^{n+1} - u_j^n}{\Delta t} + O(\Delta t). \tag{4.79}$$

This is called *forward Euler* differencing. While forward Euler differencing is only first-order accurate in Δt, it has the advantage that one is able to calculate quantities at timestep $n+1$ in terms of only quantities known at time-step n. For the space derivative, we can use a second-order representation still using only quantities known at time-step n

$$\left.\frac{\partial u}{\partial t}\right|_{j,n} = \frac{u_{j+1}^n - u_{j-1}^n}{2\Delta x} + O(\Delta x^2). \tag{4.80}$$

The resulting finite-difference approximation to Eq. (4.76) is called the FTCS representation (Forward Time Centered Space)

$$\frac{u_j^{n+1} - u_j^n}{\Delta t} = -v \left(\frac{u_{j+1}^n - u_{j-1}^n}{2\Delta x} \right), \tag{4.81}$$

which can easily be rearranged to be a formula for u_j^{n+1} in terms of the other quantities. The FTCS scheme is a fine example of an algorithm that is easy to derive, takes little storage, and executes quickly. Unfortunately it does not work!

The FTCS representation is an *explicit* scheme. This means that u_j^{n+1} for each j can be calculated explicitly from the quantities that are already known. Later we shall meet *implicit* schemes, which require us to solve implicit equations coupling the u_j^{n+1} for various j. The FTCS algorithm is also an example of a *single-level* scheme, since only values at time level n have to be stored to find values at time level $n+1$.

4.4.1   Von Neumann Stability Analysis

Unfortunately, Eq. (4.81) is of very limited usefulness. It is an *unstable* method, which can be used only (if at all) to study waves for a short fraction of one oscillation period. To find alternative methods with more general applicability, we must introduce the *von Neumann stability analysis*. The von Neumann analysis is local: We imagine that the coefficients of the difference equations are so slowly varying as to be considered constant in space and time. In that case, the independent solutions, or *eigenmodes*, of the difference equations are all of the form

$$u_j^n = \xi^n e^{ikj\Delta x}, \tag{4.82}$$

where k is a real spatial wave number (which can have any value) and $\xi = \xi(k)$ is a complex number that depends on k. The key fact is that the time dependence of a single eigenmode is nothing more than successive integer powers of the complex number ξ. Therefore, the difference equations are unstable (have exponentially growing modes) if $|\xi(k)| > 1$ for *some* k. The number ξ is called the *amplification factor* at a given wave number k. To find $\xi(k)$, we simply substitute Eq. (4.82) back into Eq. (4.81). Dividing by ξ^n, we get

$$\xi(k) = 1 - i\frac{v\Delta t}{\Delta x}\sin(k\Delta x), \tag{4.83}$$

whose modulus is >1 for *all* k; so the FTCS scheme is unconditionally unstable. If the velocity v were a function of t and x, then we would write v_j^n in Eq. (4.81). In the von Neumann stability analysis we would still treat v as a constant, the idea being that for v slowly varying, the analysis is local. In fact, even in the case of strictly constant v, the von Neumann analysis does not rigorously treat the end effects at $j = 0$ and $j = N$. More generally, if the equation's right-hand side was nonlinear in u, then a von Neumann analysis would linearize this nonlinearity by writing $u = u_0 + \delta u$, expanding to linear order in δu. Assuming that the u_0 quantities already satisfy the difference equation exactly, the analysis would look for an unstable eigenmode of δu. Despite its lack of rigor, the von Neumann method generally gives valid answers and is much easier to apply than more careful methods. We accordingly adopt it exclusively. (See, for example, [68] for a discussion of other methods of stability analysis.)

4.4.2 Lax Method

The instability in the FTCS method can be cured by a simple change due to Lax. One replaces the term u_j^n in the time derivative term by its average

$$u_j^n \rightarrow \frac{1}{2}\left(u_{j+1}^n + u_{j-1}^n\right) \tag{4.84}$$

This turns Eq. (4.81) into

$$u_j^{n+1} = \frac{1}{2}\left(u_{j+1}^n - u_{j-1}^n\right) - \frac{v\Delta t}{2\Delta x}\left(u_{j+1}^n - u_{j-1}^n\right). \tag{4.85}$$

Substituting Eq. (4.82), we find for the amplification factor

$$\xi = \cos k\Delta x - i\frac{v\Delta t}{\Delta x}\sin k\Delta x. \tag{4.86}$$

The stability condition $|\xi|^2 \leq 1$ leads to the requirement

$$\frac{|v|\,\Delta t}{\Delta x} \leq 1. \tag{4.87}$$

This is the famous Courant–Friedrichs–Lewy stability criterion, often called simply the *Courant condition*. Intuitively, the stability condition can be understood as follows: The quantity u_j^{n+1} in Eq. (4.85) is computed from information at points $j - 1$ and $j + 1$ at time n. In other words, x_{j-1} and x_{j+1} are the boundaries of the spatial region that is allowed to communicate information to u_j^{n+1}. Now recall that in the continuum wave equation, information actually propagates with a maximum velocity v. If the point u_j^{n+1} is outside of the shaded region, then it requires information from points more distant than the differencing scheme allows. Lack of that information gives rise to an instability. Therefore, Δt cannot be made too large. The surprising result, that the simple replacement Eq. (4.84) stabilizes the FTCS scheme, is our first encounter with the fact that differencing PDEs is an art as much as a science. To see if we can demystify the art somewhat, let us compare the FTCS and Lax schemes by rewriting Eq. (4.85) so that it is in the form of Eq. (4.81) with a remainder term

$$\frac{u_j^{n+1} - u_j^n}{\Delta t} = -v\left(\frac{u_{j+1}^n - u_{j-1}^n}{2\Delta x}\right) + \frac{1}{2}\left(\frac{u_{j+1}^n - 2u_j^n + u_{j-1}^n}{\Delta t}\right). \tag{4.88}$$

However, this is exactly the FTCS representation of the equation

$$\frac{\partial u}{\partial t} = -v\frac{\partial u}{\partial x} + \frac{(\Delta x)^2}{2\Delta t}\nabla^2 u, \tag{4.89}$$

where $\nabla^2 = \partial^2/\partial x^2$ in one dimension. We have, in effect, added a diffusion term to the equation, or a dissipative term. The Lax scheme is thus said to have *numerical dissipation*, or *numerical viscosity*. We can see this also in the amplification factor. Unless $|v|\Delta t$ is exactly equal to Δx, $|\xi| < 1$ and the amplitude of the wave decreases spuriously. Is not a spurious decrease as bad as a spurious increase? The answer is no. The scales that we hope to study accurately are those that encompass many grid points, so that they have $k\Delta x \sim 1$. For these scales, the amplification factor can be seen to be very close to one, in both the stable and unstable schemes. The stable and unstable schemes are therefore about equally accurate. For the unstable scheme, however, short scales with $k\Delta x \sim 1$, *which we are not interested in*, will blow up and swamp the interesting part of the solution. It is much better to have a stable scheme in which these short wavelengths die away innocuously. Both the stable and the unstable schemes are *inaccurate* for these short wavelengths, but the inaccuracy is of a tolerable character when the scheme is stable.

4.4.3 Other Varieties of Error

Thus far we have been concerned with *amplitude error*, because of its intimate connection with the stability or instability of a differencing scheme. Other varieties of error are relevant when we shift our concern to accuracy, rather than stability. Finite-difference schemes for hyperbolic equations can exhibit dispersion, or *phase errors*. For example, Eq. (4.86) can be rewritten as

$$\xi = e^{-ik\Delta x} + i \left(1 - \frac{v\Delta t}{\Delta x} \right) \sin k\Delta x. \tag{4.90}$$

An arbitrary initial wave packet is a superposition of modes with different k's. At each timestep the modes get multiplied by different phase factors given in Eq. (4.90), depending on their value of k. If $\Delta t = \Delta x/v$, then the exact solution for each mode of a wave packet $f(x - vt)$ is obtained if each mode gets multiplied by $\exp(-ik\Delta x)$. For this value of Δt, Eq. (4.90) shows that the finite-difference solution gives the exact analytic result. However, if $v\Delta t/\Delta x$ is not exactly 1, the phase relations of the modes can become hopelessly garbled and the wave packet disperses. Note from Eq. (4.90) that the dispersion becomes large as soon as the wavelength becomes comparable to the grid spacing Δx.

A third type of error is one associated with nonlinear hyperbolic equations and is therefore sometimes called *nonlinear instability*. For example, a piece of the Euler or Navier–Stokes equations for fluid flow looks like

$$\frac{\partial v}{\partial t} = -v\frac{\partial v}{\partial x} + \dots \tag{4.91}$$

The nonlinear term in v can cause a transfer of energy in Fourier space from long wavelengths to short wavelengths. This results in a wave profile steepening until a vertical profile or "shock" develops. Since the von Neumann analysis suggests that the stability can depend on $k\Delta x$, a scheme that was stable for shallow profiles can become unstable for steep profiles. This kind of difficulty arises in a differencing scheme where the cascade in Fourier space is halted at the shortest wavelength representable on the grid, that is, at $k \sim 1/\Delta x$. If energy simply accumulates in these modes, it eventually swamps the energy in the long wavelength modes of interest. Nonlinear instability and shock formation is thus somewhat controlled by numerical viscosity such as that discussed in connection with Eq. (4.88) above. In some fluid problems, however, shock formation is not merely an annoyance, but an actual physical behavior of the fluid whose detailed study is a goal. Then, numerical viscosity alone may not be adequate or sufficiently controllable. This is a complicated subject which we discuss further in the subsection on fluid dynamics, below. For wave equations, propagation errors (amplitude or phase) are usually most worrisome. For advective equations, on the other hand, *transport errors* are usually of greater concern. In the Lax scheme, Eq. (4.85), a disturbance in the advected quantity u at mesh point j propagates to mesh points $j + 1$ and $j - 1$ at the next timestep. In reality, however, if the velocity v is positive, then only mesh point $j + 1$ should be affected. The simplest way to model the transport properties "better" is to use *upwind differencing*:

$$\frac{u_j^{n+1} - u_j^n}{\Delta t} = -v_j^n \begin{cases} \dfrac{u_j^n - u_{j-1}^n}{\Delta t}, & v_j^n > 0 \\[2mm] \dfrac{u_{j+1}^n - u_j^n}{\Delta t}, & v_j^n < 0 \end{cases}. \qquad (4.92)$$

Note that this scheme is only first order, not second order, accurate in the calculation of the spatial derivatives. How can it be "better"? The answer is one that annoys the mathematicians: The goal of numerical simulations is not always "accuracy" in a strictly mathematical sense, but sometimes "fidelity" to the underlying physics in a sense that is looser and more pragmatic. In such contexts, some kinds of error are much more tolerable than others. Upwind differencing generally adds fidelity to problems where the advected variables are liable to undergo sudden changes of state, e.g., as they pass through shocks or other discontinuities. One has to be guided by the specific nature of a specific problem. For the differencing scheme Eq. (4.92), the amplification factor (for constant v) is

$$\xi = 1 - \left| \frac{v \Delta t}{\Delta x} \right| (1 - \cos k \Delta x) - i \frac{v \Delta t}{\Delta x} \sin k \Delta x, \qquad (4.93)$$

$$|\xi|^2 = 1 - 2 \left| \frac{v \Delta t}{\Delta x} \right| \left(1 - \left| \frac{v \Delta t}{\Delta x} \right| \right) (1 - \cos k \Delta x). \qquad (4.94)$$

So the stability criterion $|\xi|^2 \leq 1$ is (again) simply the Courant condition given in Eq. (4.87). There are various ways of improving the accuracy of first-order upwind differencing. In the continuum equation, material originally a distance $v \Delta t$ away arrives at a given point after a time interval Δt. In the first-order method, the material always arrives from Δx away. If $v \Delta t \ll \Delta x$ (to insure accuracy), this can cause a large error. One way of reducing this error is to interpolate u between $j - 1$ and j before transporting it. This gives effectively a second-order method. Various schemes for second-order upwind differencing are discussed and compared in [69, 70].

4.4.4   Second-Order Accuracy in Time

When using a method that is first-order accurate in time but second-order accurate in space, one generally has to take $v \Delta t$ significantly smaller than Δx to achieve desired accuracy, say, by at least a factor of 5. Thus, the Courant condition is not actually the limiting factor with such schemes in practice. However, there are schemes that are second-order accurate in both space and time, and these can often be pushed right to their stability limit, with correspondingly smaller computation times. For example, the *staggered leapfrog* method for the conservation

Eq. (4.75) is defined as follows: Using the values of u^n at time t_n, compute the fluxes F_j^n. Then compute new values u^{n+1} using the time-centered values of the fluxes:

$$u_j^{n+1} - u_j^{n-1} = -\frac{\Delta t}{\Delta x} \left(F_{j+1}^n - F_{j-1}^n \right). \tag{4.95}$$

The name comes from the fact that the time levels in the time derivative term "leapfrog" over the time levels in the space derivative term. The method requires that u^{n-1} and u^n be stored to compute u^{n+1}. For our simple model Eq. (4.76), the staggered leapfrog approach takes the form

$$u_j^{n+1} - u_j^{n-1} = -\frac{v\Delta t}{\Delta x} \left(u_{j+1}^n - u_{j-1}^n \right). \tag{4.96}$$

The von Neumann stability analysis now gives a quadratic equation for ξ, rather than a linear one, because of the occurrence of three consecutive powers of ξ when the form given in Eq. (4.82) for an eigenmode is substituted into Eq. (4.96)

$$\xi^2 - 1 = -2i\xi \frac{v\Delta t}{\Delta x} \sin k\Delta x, \tag{4.97}$$

whose solution is

$$\xi = -i\frac{v\Delta t}{\Delta x} \sin k\Delta x \pm \sqrt{1 - \left(\frac{v\Delta t}{\Delta x} \sin k\Delta x \right)^2}. \tag{4.98}$$

Thus, the Courant condition is again required for stability. In fact, in Eq. (4.98), $|\xi|^2 = 1$ for any $v\Delta t \leq \Delta x$. This is the great advantage of the staggered leapfrog method: There is no amplitude dissipation. Staggered leapfrog differencing of equations is most transparent if the variables are centered on appropriate half-mesh points:

$$r_{j+1/2}^n = v \left. \frac{\partial u}{\partial x} \right|_{j+1/2}^n = v\frac{u_{j+1}^n - u_j^n}{\Delta x},$$

$$s_j^{n+1/2} = v \left. \frac{\partial u}{\partial x} \right|_j^{n+1/2} = v\frac{u_j^{n+1} - u_j^n}{\Delta t}. \tag{4.99}$$

This is purely a notational convenience: we can think of the mesh on which r and s are defined as being twice as fine as the mesh on which the original variable u is defined. The leapfrog differencing is

$$\frac{r_{j+1/2}^{n+1} - r_{j+1/2}^n}{\Delta t} = \frac{s_{j+1}^{n+1/2} - s_j^{n+1/2}}{\Delta x},$$

$$\frac{s_j^{n+1/2} - s_j^{n-1/2}}{\Delta t} = v\frac{r_{j+1/2}^n - r_{j-1/2}^n}{\Delta x}. \tag{4.100}$$

If you substitute in Eq. (4.100), you will find that once again the Courant condition is required for stability, and that there is no amplitude dissipation when it is satisfied. If we substitute Eq. (4.99) in Eq. (4.100), we find that Eq. (4.100) is equivalent to

$$\frac{u_j^{n+1} - 2u_j^n + u_j^{n-1}}{(\Delta t)^2} = v^2 \frac{u_{j+1}^n - 2u_j^n + u_{j-1}^n}{(\Delta x)^2}. \tag{4.101}$$

This is just the "usual" second-order differencing of the wave equation. We see that it is a two-level scheme, requiring both u^n and u^{n-1} to obtain u^{n+1}. In Eq. (4.100) this shows up as both $s^{n-1/2}$ and r^n being needed to advance the solution.

For equations more complicated than our simple model equation, especially nonlinear equations, the leapfrog method usually becomes unstable when the gradients get large. The instability is related to the fact that odd and even mesh points are completely decoupled. This mesh drifting instability is cured by coupling the two meshes through a numerical viscosity term, e.g., adding to the right side of Eq. (4.96) a small coefficient ($\ll 1$) times $u_{j+1}^n - 2u_j^n + u_{j-1}^n$. For more on stabilizing difference schemes by adding numerical dissipation, see, e.g., [71].

The *Two-Step Lax–Wendroff* scheme is a second order in time method that avoids large numerical dissipation and mesh drifting. One defines intermediate values $u_{j+1/2}$ at the half timesteps $t_{n+1/2}$ and the half mesh points $x_{j+1/2}$. These are calculated by the Lax scheme:

$$u_{j+1/2}^{n+1/2} = \frac{1}{2}\left(u_{j+1}^n + u_j^n\right) - \frac{\Delta t}{2\Delta x}\left(F_{j+1}^n - F_j^n\right). \tag{4.102}$$

Using these variables, one calculates the fluxes $F_{j+1/2}^{n+1/2}$. Then the updated values u_j^{n+1} are calculated by the properly centered expression

$$u_j^{n+1} = u_j^n - \frac{\Delta t}{\Delta x}\left(F_{j+1/2}^{n+1/2} - F_{j-1/2}^{n+1/2}\right). \tag{4.103}$$

The provisional values $u_{j+1/2}^{n+1/2}$ are now discarded. Let us investigate the stability of this method for our model advective equation, where $F = vu$. Denoting $\alpha = \frac{v\Delta t}{\Delta x}$, we get

$$\xi = 1 - i\alpha \sin k\Delta x - \alpha^2 (1 - \cos k\Delta x) \tag{4.104}$$

so

$$|\xi|^2 = 1 - \alpha^2\left(1 - \alpha^2\right)(1 - \cos k\Delta x)^2 \tag{4.105}$$

The stability criterion $|\xi|^2 \le 1$ is therefore $\alpha^2 \le 1$, or $v\Delta t \le \Delta x$ as usual. Incidentally, one should not think that the Courant condition is the only stability requirement that ever turns up in PDEs. It keeps doing so in our model examples just because those examples are so simple in form. The method of analysis is, however, general. Except when $\alpha = 1$, $|\xi|^2 < 1$ in Eq. (4.105), so some amplitude damping does occur. The effect is relatively small, however, for wavelengths large compared with the mesh size Δx. If we expand Eq. (4.105) for small $k\Delta x$, we find

$$|\xi|^2 = 1 - \alpha^2(1 - \alpha^2)\frac{(k\Delta x)^4}{4} + \dots . \qquad (4.106)$$

The departure from unity occurs only at fourth order in k. This should be contrasted with Eq. (4.86) for the Lax method, which shows that

$$|\xi|^2 = 1 - (1 - \alpha^2)(k\Delta x)^2 + \dots \qquad (4.107)$$

for small $k\Delta x$. In summary, our recommendation for initial value problems that can be cast in flux-conservative form, and especially problems related to the wave equation, is to use the staggered leapfrog method when possible. We have personally had better success with it than with the Two-Step Lax–Wendroff method. For problems sensitive to transport errors, upwind differencing or one of its refinements should be considered.

4.4.5   Fluid Dynamics with Shocks

As we alluded earlier, the treatment of fluid dynamics problems with shocks has become a very complicated and very sophisticated subject. All we can attempt to do here is to provide a guide to some starting points in the literature. There are basically three important general methods for handling shocks.

The oldest and simplest method, invented by von Neumann and Richtmyer, is to add *artificial viscosity* to the equations, modeling the way Nature uses real viscosity to smooth discontinuities.

The second method combines a high-order differencing scheme that is accurate for smooth flows with a low-order scheme that is very dissipative and can smooth the shocks. Typically, various upwind differencing schemes are combined using weights chosen to zero the low-order scheme unless steep gradients are present, and also chosen to enforce various "monotonicity" constraints that prevent nonphysical oscillations from appearing in the numerical solution. Reference [72] is a good place to start with these methods.

The third, and potentially most powerful method, is Godunov's approach. Here one gives up the simple linearization inherent in finite differencing based on Taylor series and includes the nonlinearity of the equations explicitly. There is an analytic solution for the evolution of two uniform states of a fluid separated by a discontinuity, the Riemann shock problem. Godunov's idea was to approximate the fluid by a large number of cells of uniform states, and piece them together using the Riemann solution. There have been many generalizations of Godunov's approach, of which the most powerful is probably the PPM method [73].

Readable reviews of all these methods, discussing the difficulties arising when one-dimensional methods are generalized to multiple dimensions, are given in [74–76].

PROBLEMS FOR CHAPTER 4:

1. In almost all derivations one assumes that deformation potential scattering is an elastic process, which is not exactly true. It is, therefore, interesting to see what the average energy loss per unit time of a carrier to the crystal lattice actually is. The average energy loss per unit time per carrier is defined as

$$\left\langle \frac{dE}{dt} \right\rangle_{coll} = \frac{\int f(\mathbf{k})\,(\partial E/\partial t)_{coll}\,d^3k}{\int f(\mathbf{k})d^3k},$$

where the energy relaxation rate $(\partial E/\partial t)_{coll}$ is given by

$$(\partial E/\partial t)_{coll} = \sum_{\mathbf{k}'} \left[E(\mathbf{k}) - E(\mathbf{k}')\right] S(\mathbf{k}, \mathbf{k}') \left[1 - f(\mathbf{k}')\right].$$

In the above expression, one must consider both the absorption and the emission process. Show that the average energy loss per unit time of a carrier for a degenerate electron gas is given by

$$\left\langle \frac{dE}{dt} \right\rangle_{coll} = \frac{2m^* \Xi_{ac}^2}{\pi^{3/2} \rho \hbar} \left(\frac{2m^* k_B T_e}{\hbar^2}\right)^{3/2} 2\frac{T_e - T_L}{T_e}\frac{\Im_1(\eta)}{\Im_{1/2}(\eta)},$$

where $\Im_i(\eta)$ are the Fermi–Dirac integrals, $\eta = E_F/k_B T_e$ is the reduced Fermi energy, T_e is the electron and T_L is the lattice temperature. (Hints: Where appropriate, assume that acoustic phonon scattering is nearly elastic scattering process. Also, $df/dE = f(f-1)/k_B T_e$ is a useful identity).

2. Derive the expression for the momentum relaxation time appearing in the HD equations for simple acoustic phonon scattering and for Coulomb scattering under the assumption that Coulomb scattering is both not screened and is strongly screened.

3. Develop a one-dimensional (1D) hydro-dynamic simulator for modeling pn-junctions (diodes) under forward and reverse bias conditions. Include both types of carriers in your model (electrons and holes).

   Model:
   Silicon diode, with permittivity $\varepsilon_{sc} = 1.05 \times 10^{-10}\,\mathrm{Fm}^{-1}$ and intrinsic carrier concentration $n_i = 1.5 \times 10^{10}\,\mathrm{cm}^{-3}$ at $T = 300$ K. In all your simulations assume that $T = 300$ K. Use concentration-dependent and field-dependent mobility models and SRH generation–recombination process. Assume ohmic contacts and charge neutrality

at both ends to get the appropriate boundary conditions for the potential and the electron and hole concentrations. For the concentration-dependent electron and hole mobilities use the Arora model given in the Silvaco Manuals. For the field-dependent mobility, use the model described in the Silvaco ATLAS manual, using the relevant parameters listed there. For the SRH generation–recombination, use the expressions given in the text. To simplify your calculations, assume that the trap energy level coincides with the intrinsic level. Use momentum and energy relaxation times in the range of values used in the literature.

Doping:
Use $N_A = 10^{16}$ cm^{-3} and $N_D = 10^{17}$ cm^{-3} as a net doping of the p- and n-regions, respectively.

Outputs:

- Vary the anode bias V_A from 0 to 0.6 V, in voltage increments that are fraction of the thermal voltage $V_T = k_B T/q$, to have stable convergence. Plot the resulting IV characteristics. Compare these results with the results that you have obtained with the DD model in a previous exercise.

- For anode bias of 0.6 V plot the velocity of the carriers along the x-axis for the different values that you have used for the momentum and energy relaxation times. Under what conditions these velocity plots approach the DD results? What effect you have included in the model? Explain this with physical reasoning.

CHAPTER 5

Use of Commercially Available Device Simulators

5.1 THE NEED FOR SEMICONDUCTOR DEVICE MODELING

At present there are several trends in the semiconductor industry that are occurring concurrently with rapid changes in the applications of semiconductors. The competitiveness among many semiconductor manufacturers is shifting from an emphasis on technology and fabrication to a much greater emphasis on product design, architecture, algorithm, and software; i.e., shifting from technology-oriented R&D to product-oriented R&D in which computers, modeling, and simulation become increasingly crucial for marketplace success. Other trends include:

- Increased costs for R&D and production facilities, which are becoming too large for any one company or country to sustain.
- Shorter process technology life cycles.
- Emphasis on faster characterization of manufacturing processes, assisted by modeling and simulation.

Computer simulations, often called technology for computer-aided design (TCAD) offer many advantages such as:

- Evaluating "what-if" scenarios rapidly
- Providing problem diagnostics
- Providing full-field, in-depth understanding
- Providing insight into extremely complex problems/phenomena/product sets
- Decreasing design cycle time (savings on hardware build lead-time, gain insight for next product/process)
- Shortening time to market

5.1.1   Importance of Semiconductor Device Simulators

Computer simulations are particularly useful in meeting the demands imposed by the major industry trends identified in technology roadmaps such as the International Technology Roadmap for Semiconductors (*ITRS*), the National Electronics Manufacturing Initiative (*NEMI*), and the Optoelectronics Industry Development Association (*OIDA*); namely—lower product cycle time, increasing product complexity at both the component and system level, "doing it right the first time," and rapid volume ramp-ups.

Some TCAD prerequisites are:

- Modeling and simulation require enormous technical depth and expertise not only in simulation techniques and tools but also in the fields of physics and chemistry.

- Laboratory infrastructure and experimental expertise are essential for both model verification and input parameter evaluations in order to have truly effective and predictive simulations.

- Software and tool vendors need to be closely tied to development activities in the research and development laboratories.

These prerequisites may have considerable business cost, confidentiality, and logistical implications, and must be carefully evaluated.

5.1.2   Key Elements of Physical Device Simulation

Physically based device simulation is not a familiar concept for all engineers. A brief overview is provided here to serve as a high-level orientation. Physically based device simulators predict the electrical characteristics that are associated with specified physical structures and bias conditions. This is achieved by approximating the operation of a device onto a two- or three-dimensional grid, consisting of a number of grid points called nodes. By applying a set of differential equations, derived from Maxwells laws, onto this grid it is possible to simulate the transport of carriers through a structure. This means that the electrical performance of a device can now be modeled in DC, AC, or transient modes of operation. Physically based simulation provides three major advantages: it is predictive, it provides insight, and it captures theoretical knowledge in a way that makes this knowledge available to nonexperts. Physically based simulation is different from empirical modeling. The goal of empirical modeling is to obtain analytic formulae that approximate existing data with good accuracy and minimum complexity. Empirical models provide efficient approximation and interpolation. They do not usually provide insight, or predictive capabilities, or encapsulation of theoretical knowledge. Physically based simulation is an alternative to experiments as a source of data. Physically based simulation has become very important for two reasons. First, it is almost always much quicker and cheaper than performing experiments. Second, it provides information that is difficult or impossible to measure. The drawbacks of simulation are that all the relevant physics must be incorporated into a simulator, and numerical procedures must be implemented to solve the associated equations. Users of physically based device simulation tools must specify the problem to be simulated. Users of ATLAS, for example, specify device simulation problems by defining:

- The physical structure to be simulated
- The physical models to be used
- The bias conditions for which electrical characteristics are to be simulated.

5.1.3   Historical Development of the Physical Device Modeling

As already noted in the previous section, along with the technological discoveries came theories which using closed form analytical solutions to explain the operation of, for example, the MOSFET using the gradual channel approximation. The Schockley model for describing the operation of *pn*-junctions and BJTs was also very successful in explaining the corresponding experimental characteristics of these devices. It was not until 1964 when numerical modeling started to play some role in the design and development of experimental devices. At that time, Gummel [48] introduced the decoupled scheme for the solution of the Poisson and the continuity equations for a BJT. If, for example, one chooses the quasi-Fermi level formulation discussed in Chapter 2, one first solves the nonlinear Poisson's equation. The potential obtained is substituted into the continuity equations, which are now linear, and are solved directly to conclude the iteration. The result in terms of quasi-Fermi levels is then substituted back into Poisson's equation until convergence is reached. In order to check for convergence, one can calculate the residuals obtained by positioning all the terms to the left-hand side of the equations and substituting the variables with the iteration values, as discussed in Chapter 3. For exact solution values, the residuals should be zero. Convergence is assumed when the residuals are smaller than a set tolerance. The rate of convergence of the Gummel method is faster when there is little coupling between the different equations. In 1968 [51] in his investigation of PN diodes, de Mari introduced the scaling of variables that is used even today and effectively prevents overflows and underflows that occur during the simulation sequence. Even though these two previous efforts could successfully model *pn*-junctions and BJT's, they suffered from the problem of negative densities that prevent the convergence of the code because a simple finite difference scheme was used for the discretization of the continuity equations. In particular, it was assumed that both the potential and the electron densities vary linearly in between mesh nodes. This is only the case when an infinitely small mesh is used which, on the other hand, increases the number of unknowns and the sizes of the matrices that need to be inverted for a proper solution to be obtained. A breakthrough that overcame this problem occurred in 1969 when Sharfetter and Gummel, in their seminal paper that describes the simulation of a 1D Silicon Read (IMPATT) diode, introduced the so-called Sharfetter–Gummel discretization of the continuity equation [7]. In other words, Sharfetter and Gummel assumed that the potential can be a linearly-varying function between mesh points. However, since the dependence of the electron density on the potential is exponential, the electron density must preserve this exponential dependence between mesh points. This fact allowed them to use much larger mesh sizes, and the convergence of the Gummel iteration scheme was significantly improved. The introduction of the Sharfetter–Gummel discretization scheme allowed numerous 2D/3D device simulators to be developed, among which we note the Kennedy and O'Brien work in 1970 on 2D simulations of silicon JFETs [77], Slotboom's 2D simulation of BJT's in 1973 [78], and the Yoshii *et al.* 3D modeling of a range of semiconductor devices in 1982 [79].

With industry recognizing the need for physical device simulation for shortening the design to production process, a number of commercial device simulators appeared. These are classified below based on the capability of simulating a particular device technology:

2D MOS:	MINIMOS, GEMINI, PISCES, CADDET, HFIELDS, CURRY
3D MOS:	WATMOS, FIELDAY, MINIMOS3D
1D BJT:	SEDAN, BIPOLE, LUSTRE
2D BJT:	BAMBI, CURRY
MESFETs:	CUPID

The above-listed simulators are presently used on an everyday basis in the optimization of the electrical characteristics of the devices for a given doping profiles, device layout, etc. However, as the devices are scaled deeper to submicron dimensions, new physical phenomena start to appear which cannot be captured by the physics of drift–diffusion (DD) or even the energy balance models discussed in Chapters 3 and 4. Hence simulators that rely on the direct solution of the Boltzmann transport equation have been developed by various groups around the world, the most well-established being the DAMOCLES simulator developed by Massimo Fischetti and Stephen Laux at IBM T. J. Watson Institute in Yorktown Heights [80]. Even though it goes beyond the scope of this book which is limited to semiclassical device modeling, it is worthwhile to mention one more publically available simulation package (NEMO1D) based on the solution of a fully quantum mechanical recursive Green's function method that has shown to be very useful for modeling one-dimensional quantum structures such as resonant tunneling diodes [13]. The major advantages and limitations of the most commonly used semiclassical simulation tools are listed in Figure 5.1.

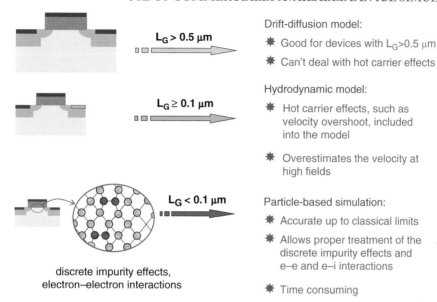

Drift-diffusion model:

* Good for devices with $L_G > 0.5$ µm
* Can't deal with hot carrier effects

Hydrodynamic model:

* Hot carrier effects, such as velocity overshoot, included into the model

* Overestimates the velocity at high fields

Particle-based simulation:

* Accurate up to classical limits
* Allows proper treatment of the discrete impurity effects and e–e and e–i interactions

* Time consuming

$L_G > 0.5$ µm

$L_G \geq 0.1$ µm

$L_G < 0.1$ µm

discrete impurity effects, electron–electron interactions

FIGURE 5.1: Limitations and advantages of some of the semiclassical simulation tools in capturing the proper device physics

5.2   INTRODUCTION TO THE SILVACO ATLAS SIMULATION TOOL

ATLAS is a modular and extensible framework for one-, two-, and three-dimensional semiconductor device simulation [81]. ATLAS should only be used with Virtual Wafer Fab (VWF) Interactive Tools. These include DECKBUILD, TONYPLOT, DEVEDIT, MASKVIEWS, and OPTIMIZE. DECKBUILD provides an interactive run-time environment. TONYPLOT supplies scientific visualization capabilities. DEVEDIT is an interactive tool for structure and mesh specification and refinement, and MASKVIEWS is an IC Layout Editor. The OPTIMIZER supports blackbox optimization across multiple simulators. ATLAS is very often used in conjunction with the ATHENA process simulator. ATHENA predicts the physical structures that result from processing steps. The resulting physical structures are used as input by ATLAS, which then predicts the electrical characteristics associated with specified bias conditions. The combination of ATHENA and ATLAS makes it possible to determine the impact of process parameters on device characteristics.

Figure 5.2 shows the types of information that flow in and out of ATLAS. Most ATLAS simulations use two inputs: a text file that contains commands for ATLAS to execute, and a structure file that defines the structure that will be simulated. ATLAS produces three types

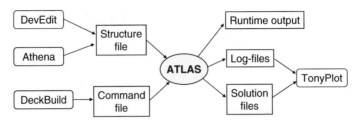

FIGURE 5.2: ATLAS inputs and outputs

of output. The run-time output provides a guide to the progress of simulations running, and is where error messages and warning messages appear. Log files store all terminal voltages and currents from the device analysis, and solution files store two- and three-dimensional data relating to the values of solution variables within the device for a single bias point.

5.2.1   The ATLAS Syntax

An ATLAS command file is a list of commands for ATLAS to execute. This list is stored as an ASCII text file that can be prepared in DECKBUILD or using any text editor. Preparation of the input file in DECKBUILD is preferred, and can be made easier by appropriate use of the DECKBUILD Commands menu. The input file contains a sequence of statements. Each statement consists of a keyword that identifies the statement and a set of parameters. The general format is:

```
<STATEMENT> <PARAMETER>=<VALUE>
```

Some hints on the proper structure of the statements are listed below:

1. The statement keyword must come first, but after this the order of parameters within a statement is not important.

2. It is only necessary to use enough letters of any parameter to distinguish it from any other parameter on the same statement. Thus, CONCENTRATION can be shortened to CONC. However, REGION cannot be shortened to R since there is also a parameter RATIO associated with the DOPING statement.

3. Logicals can be explicitly set to false by preceding them with the ^ symbol.

4. Any line beginning with # is ignored. These lines are used as comments.

5. ATLAS can read up to 256 characters on one line. However, it is best to spread long input statements over several lines to make the input file more readable. The character \ at the end of a line indicates continuation.

Group		Statements
1. Structure specification	_____	MESH REGION ELECTRODE DOPING
2. Material models specification	_____	MATERIAL MODELS CONTACT INTERFACE
3. Numerical method selection	_____	METHOD
4. Solution specification	_____	LOG SOLVE LOAD SAVE
5. Results analysis	_____	EXTRACT TONYPLOT

FIGURE 5.3: ATLAS command groups with the primary statements in each group

The order in which statements occur in an ATLAS input file is important. There are five groups of statements, and these must occur in the correct order. These groups are indicated in Figure 5.3. *Each input file must contain these five groups in order.* Failure to do this will usually cause an error message and termination of the program, but it could also lead to incorrect operation of the program. For example, material parameters or models set in the wrong order may not be used in the calculations. The order of statements within the mesh definition, structural definition, and solution groups is also important.

A device structure can be defined in three different ways in ATLAS:

1. An existing structure can be read in from a file. The structure can be created by earlier ATLAS run or by another program such as ATHENA or DEVEDIT. A single statement loads in the mesh, geometry, electrode positions, and DOPING of the structure. This statement is: MESH INFILE=<filename>

2. The input structure can be transferred from ATHENA or DEVEDIT through the automatic interface feature of DECKBUILD.

3. A structure can be constructed using the ATLAS command language.

4. The first and second methods are more convenient than the third and are to be preferred whenever possible.

5.2.2 Choice of the Numerical Method

Several different numerical methods can be used for calculating the solutions of semiconductor device problems. Different solution methods are optimum in different situations and some guidelines are given here. Different combinations of models may require ATLAS to solve up to six equations. For each of the model types, there are basically three types of solution techniques: (a) decoupled (GUMMEL), (b) fully coupled (NEWTON), and (c) BLOCK. In simple terms, a decoupled technique like the Gummel method will solve for each unknown in turn, keeping the other variables constant, repeating the process until a stable solution is achieved. Fully coupled techniques, such as the Newton method, solve the total system of unknowns together. The combined or block methods will solve some equations fully coupled, while others are decoupled.

In general, the Gummel method is useful where the system of equations is weakly coupled, but has only linear convergence. The Newton method is useful when the system of equations is strongly coupled and has quadratic convergence. The Newton method may however spend extra time solving for quantities, which are essentially constant or weakly coupled. Newton also requires a more accurate initial guess to the problem to obtain convergence. Thus, a block method can provide for faster simulation times in these cases over Newton. Gummel can often provide a better initial guess to problems. It can be useful to start a solution with a few Gummel iterations to generate a better guess, and then switch to Newton's method to complete the solution. Specification of the solution method is carried out as follows:

```
METHOD GUMMEL BLOCK NEWTON
```

The exact meaning of the statement depends upon the particular models it is applied to, as discussed below.

5.2.2.1 Basic Drift–Diffusion Calculations

The isothermal DD model requires the solution of three equations for the potential, the electron concentration, and the hole concentration. Specifying GUMMEL or NEWTON alone will produce simple Gummel or Newton solutions as detailed above. For almost all cases the Newton method is preferred and it is the default. Specifying:

```
METHOD GUMMEL NEWTON
```

will cause the solver to start with Gummel iterations and then switch to Newton, if convergence is not achieved. This approach is a very robust, although more time consuming way of obtaining solutions for any device. However, this method is highly recommended for all simulations with floating regions such as Si on Insulator (SOI) transistors. A floating region is defined as an area of doping which is separated from all electrodes by a *pn*-junction. BLOCK is equivalent to NEWTON for all isothermal DD simulations.

5.2.2.2 Drift–Diffusion Calculations with Lattice Heating

When the lattice-heating model is added to drift–diffusion, an extra equation is added. The BLOCK algorithm solves the three DD equations as a Newton solution and follows this with a decoupled solution of the heat flow equation. The NEWTON algorithm solves all four equations in a coupled manner. NEWTON is preferred once the temperature is high, however BLOCK is quicker for low temperature gradients. Typically the combination used is:

```
METHOD BLOCK NEWTON
```

5.2.2.3 Energy Balance Calculations

The energy balance model requires the solution of up to five coupled equations. GUMMEL and NEWTON have the same meanings as with the DD model (i.e., GUMMEL specifies a decoupled solution and NEWTON specifies a fully coupled solution). However, BLOCK performs a coupled solution of the potential and carrier continuity equations, followed by a coupled solution of the carrier energy balance, and carrier continuity equations. It is possible to switch from BLOCK to NEWTON by specifying multiple solution methods on the same line. For example:

```
METHOD BLOCK NEWTON
```

will begin with BLOCK iterations then switch to NEWTON if convergence is still not achieved. This is the most robust approach for many energy balance applications. The points at which the algorithms switch is predetermined, but can also be changed on the METHOD statement. The default values set by Silvaco work well for most circumstances.

5.2.2.4 Energy Balance Calculations with Lattice Heating

When nonisothermal solutions are performed in conjunction with energy balance models, a system of up to six equations must be solved. GUMMEL or NEWTON solve the equations iteratively or fully coupled, respectively. BLOCK initially performs the same function as with energy balance calculations; then it solves the lattice heating equation in a decoupled manner.

ATLAS can solve both the electron and hole continuity equations, or only for one or none. This choice can be made using the parameter CARRIERS. For example,

```
METHOD CARRIERS=2
```

specifies that a solution for both carriers is required. This value is the default. With one carrier the parameter ELEC or HOLE is needed. For example, for hole solutions only one uses:

```
METHOD CARRIERS=1 HOLE
```

To select a solution for the potential only specify:

```
METHOD CARRIERS=0
```

5.2.3   Solutions Obtained

ATLAS can calculate DC, AC, small signal, and transient solutions. Obtaining solutions is rather analogous to setting up parametric test equipment for device tests. The user defines the voltages on each of the electrodes in the device. ATLAS then calculates the current through each electrode. ATLAS also calculates internal quantities, such as carrier concentrations and electric fields throughout the device. This is information that is difficult or impossible to measure. In all simulations the device starts with zero bias on all electrodes. Solutions are obtained by stepping the biases on the electrodes from this initial equilibrium condition. Results are saved using the LOG or SAVE statements.

To obtain convergence for the equations used, it is necessary to supply a good initial guess for the variables to be evaluated at each bias point. The ATLAS solver uses this initial guess and iterates to a converged solution. For isothermal DD simulations, the variables are the potential and the two carrier concentrations. Provided a reasonable grid is used, almost all convergence problems in ATLAS are caused by a poor initial guess to the solution. During a bias ramp, the initial guess for any bias point is provided by a projection of the two previous results. Problems tend to appear near the beginning of the ramp when two previous results are not available. If one previous bias is available, it is used alone. The following two examples eventually produce the same result, although the first will likely have far more convergence problems than the second due to the issue of a good initial guess.

```
1. SOLVE VGATE=1.0 VDRAIN=1.0 VSUBSTRATE=-1.0
2. SOLVE VGATE=1.0
   SOLVE VSUBSTRATE=-1.0
   SOLVE VDRAIN=1.0
```

In the first case, one solution is obtained with voltages specified at all electrodes. In the second case, the solution with the gate voltage at 1.0 V is performed first. All other electrodes are at zero bias. Next, with the gate at 1.0 V, the substrate potential is raised to −1.0 V and another solution is obtained. Finally, with the substrate and the gate biased, the drain potential is added and the system solved again. The advantage of this method over the first case is that the small incremental changes in voltage allows for a better initial guesses at each step. Generally, the projection method for the initial guess gives good results when the *IV* curve is linear. However, it may encounter problems if the *IV* curve is highly nonlinear or if the device operating mode is changing. Typically this might occur around the threshold or breakdown voltages. At these biases smaller voltage steps are required to obtain convergence. ATLAS contains features such as the TRAP parameter and the, curve tracer, to automatically cut the voltage steps in these highly nonlinear areas.

Specifying AC simulations is a simple extension of the DC solution syntax. AC small-signal analysis is performed as a postprocessing operation to a DC solution. Two common types of AC simulation in ATLAS are mentioned here. These include:

1. Single-frequency AC solution during a DC ramp
2. Ramped frequency at a single bias

The results of AC simulations are the conductance and capacitance between each pair of electrodes.

5.2.4 Advanced Solution Techniques

5.2.4.1 Obtaining Solutions Around the Breakdown Voltage

Obtaining solutions around the breakdown voltage can be difficult using the standard ATLAS approach. It requires special care in the choice of voltage steps and also in interpreting the results. The curve tracer described later is the most effective method in many cases. A MOSFET breakdown simulation might be performed using this standard syntax for ramping the drain bias. Note the setting of CLIMIT as required for breakdown simulations when the prebreakdown leakage is low.

```
IMPACT SELB
METHOD CLIMIT=1e-4
SOLVE VDRAIN=1.0 VSTEP=1.0 VFINAL=20.0 NAME=drain
```

If the breakdown were 11.5 V then convergence problems will be expected for biases higher than 11.0 V using this syntax. Although technology dependent, it is common for the breakdown curve to be flat up to a voltage very close to breakdown and then almost vertical. The current changes by orders of magnitude for very small bias increments. This produces some problems for ATLAS using the syntax described above. First, if the breakdown occurs at 11.5 V, there are no solutions for voltages greater than this value. ATLAS is trying to ramp to 20.0 V so it is likely that ATLAS will fail to converge at some point. This is usually not a problem since by that point, the breakdown voltage and IV curve have been obtained. Above 11 V, bias step reduction will take place due to the TRAP parameter. ATLAS will continually try to increase the drain voltage above 11.5 V and those points will fail to converge. However, it will solve points asymptotically approaching $V_{ds} = 11.5$ V until the limit set by the MAXTRAPS parameter is reached. If the default of four traps is used it is clear that the minimum allowed voltage step is $1.0 \times (0.5)4$ or 0.004 V. This is normally enough accuracy for determining the breakdown point. However, the simulation might not allow the current to reach a sufficiently high level before MAXTRAPS is needed. Typically, in device simulation, the breakdown point is determined once the current is seen to increase above the flat prebreakdown leakage value by two orders of magnitude in a small voltage increment. If users do wish to trace the full breakdown curve up to high current values more advanced techniques than the simple voltage ramp must be used. Two of them are: curve tracer and current boundary conditions. The expense of these methods might be extra CPU time.

5.2.4.2 Using Current Boundary Conditions
In all of the examples considered in the basic description of the SOLVE statement, it was assumed that voltages were being forced and currents were being measured. ATLAS also supports the reverse case through current boundary conditions. The current through the electrode is specified in the SOLVE statement and the voltage at the contact is calculated. Current boundary conditions are set using the CONTACT statement as described earlier in this chapter. The syntax of the SOLVE statement is altered once current boundary conditions are specified.

```
SOLVE IBASE=1e-6
```

The syntax above specifies a single solution at a given current.

```
SOLVE IBASE=1e-6 ISTEP=1e-6 IFINAL=5e-6 NAME=base
```

This sets a current ramp similar in syntax to the voltage ramp described earlier.

```
SOLVE IBASE=1e-10 ISTEP=10 IMULT IFINAL=1e-6 NAME=base
```

This is similar to the previous case, but the IMULT parameter is used to specify that ISTEP should be used as a multiplier for the current rather than a linear addition. This is typical for ramps of current since linear ramps are inconvenient when several orders of magnitude in current may need to be covered. Important points to remember about current boundary conditions are that the problems of initial guess are more acute when very small (noise level) currents are used. Often it is best to ramp the voltage until the current is above 1 pA mm^{-1} and then switch to current forcing. When interpreting the results, it is important to remember the calculated voltage on the electrode with current boundary conditions is stored as the "internal bias" (e.g., base int.bias in TONYPLOT or vint. "base" in DECKBUILD's extract syntax).

5.2.4.3 The Compliance Parameter
Compliance is a parameter used to limit the current or voltage through or on an electrode during a simulation. An electrode compliance can be set and after it is reached, the bias sweep will stop. This is analogous to parametric device testing when we stop a device from being

overstressed or destroyed. The compliance refers to the maximum resultant current or voltage present after a solution is obtained. If an electrode voltage is set, then the compliance refers to the electrode current. If current boundary conditions are used, then voltage compliance can be set. The statements

```
SOLVE VGATE=1.0
SOLVE NAME=drain VDRAIN=0 VFINAL=2 VSTEP=0.2 COMPL=1E-6 CNAME=drain
```

first solves for 1 V on the gate and then ramps the drain voltage towards 2 V in 0.2 V steps. If 1 A m^{-1} of drain current is reached before $V_d = 2$ V, the simulation will stop. Thus, as in parametric testing, a particular level can be defined and the simulation can be set to solve up to that point and no further. Once the compliance limit is reached, ATLAS simulates the next statement line in the command file.

5.2.4.4 The Curve Trace Capability

The automatic curve tracing algorithm can be invoked to enable ATLAS to trace out complex *IV* curves. The algorithm can automatically switch from voltage to current boundary conditions and vice versa. A single SOLVE statement may be used to trace out complex *IV* curves such as breakdown curves and CMOS latch-up including the snapback region and second breakdown. The algorithm is based upon a dynamic load line approach. For example, typical curvetrace and solve statements to trace out an *IV* curve for the breakdown of a diode would look like:

```
CURVETRACE CONTR.NAME=cathode STEP.INIT=0.5 NEXT.RATIO=1.2 \
MINCUR=1e-12 END.VAL=1e-3 CURR.CONT
SOLVE CURVETRACE
```

The name of the electrode which is to be ramped is specified using CONTR.NAME. STEP.INIT specifies the initial voltage step. NEXT.RATIO specifies the factor used to increase the voltage step in areas on the *IV* curve away from turning points. MINCUR may be used to set a small current value above which the dynamic load line algorithm is activated. Below the MINCUR level, the STEP.INIT and NEXT.RATIO are used to determine the next solution bias. END.VAL is used to stop the tracing if the voltage or current of the ramped electrode equals or exceeds END.VAL. Either VOLT.CONT or CURR.CONT is used to specify whether END.VAL is a voltage or current value. When plotting the log file created by the curve trace statement in TONYPLOT, it is necessary to select the internal bias, labeled int.bias, for the ramped electrode instead of the plotting the applied bias, which is labelled *Voltage*.

5.2.5   Run-Time Output, Log Files, Solution Files, and the Extract Statement

As indicated in Figure 5.2, ATLAS produces three different types of output. To recap, these are:

RUN-TIME OUTPUT

This stores the run-time messages produced by ATLAS. These messages typically include important values extracted from the simulation. All error messages go to the run-time output. If a simulation fails, it is extremely important to check the run-time output for error and warning messages.

LOG FILES

Store the DC, small-signal AC, and transient terminal characteristics for a sequence of SOLVE statements. They are loaded into TONYPLOT to visualize the device behavior.

SOLUTION FILES

These store physical quantities of the structure at each grid node for a single bias point. These can be viewed in TONYPLOT to see the internal distributions of parameters (e.g., potential, electric field). They can also be loaded into other ATLAS runs to re-initialize ATLAS at nonzero biases.

5.2.5.1   Run-Time Output

Run-time output is provided in the bottom of the DECKBUILD window. If run as a batch job, the run-time output can be stored to a file. Errors occurring in the run-time output will be displayed in this window. Note that not all errors will be fatal (as DECKBUILD tries to interpret the user's file and continues). This may cause a statement to be ignored, leading to unexpected results. It is recommended that the user check the run-time output of any newly created input file, the first time it is run, to intercept any errors.

If the user specifies the PRINT option within the MODELS statement, details of material parameters, constants, and mobility models will be specified at the start of the run-time output. This is a useful way of checking what has been specified and which mobility parameters apply to which regions. It is recommended that the user always specifies MODELS PRINT in input files. During SOLVE statements the error numbers of each equation at each iteration are displayed. This is a change from the previous ATLAS version.

5.2.5.2   Log Files

Log files store the terminal characteristics calculated by ATLAS. These are current and voltages for each electrode in DC simulations. In transient simulations the time is stored and in AC simulations the small-signal frequency and the conductances and capacitances are saved. The statement:

```
LOG OUTF=<FILENAME>
```

is used to open a log file. Terminal characteristics from all SOLVE statements after the LOG statement are then saved to this file along with any results from the PROBE statement. The only way to stop the terminal characteristics being saved to this file is to use another LOG statement with either a different log filename or the parameter OFF. Typically a separate log file should be used for each bias sweep. For example, separate log files are used for each gate bias in a MOS I_d/V_{ds} simulation or each base current in a bipolar I_c/V_{ce} simulation. These files are then overlaid in TONYPLOT. Log files contain only the terminal characteristics. They are typically viewed in TONYPLOT. Parameter extraction on data in log files can be done in DECKBUILD. Log files cannot be loaded into ATLAS to reinitialize the simulation.

5.2.5.3 Parameter Extraction in DECKBUILD

The EXTRACT command is provided within the DECKBUILD environment. It allows one to extract device parameters. The command has a flexible syntax that allows you to construct very specific extract routines. EXTRACT operates on the previous solved curve or structure file. By default EXTRACT uses the currently open log file. To override this default the name of a file to be used by extract can be supplied before the extraction routine in the following way:

```
EXTRACT INIT INF="<filename>"
```

A typical example of the use of extract is the extraction of the threshold voltage of an MOS transistor. In the following example, the threshold voltage is extracted by calculating the maximum slope of the I_d/V_g curve, finding the intercept with the x-axis, and then subtracting half of the applied drain bias:

```
EXTRACT NAME="nvt" XINTERCEPT(MAXSLOPE(CURVE(V."GATE",(I."DRAIN")))\
-(AVE(V."DRAIN"))/2.0)
```

The results of the extraction will be displayed in the run-time output and will by default also be stored in the file "results.final". You can store the results in a different file by using the following option at the end of extract command:

```
EXTRACT....DATAFILE="<filename>"
```

Cutoff frequency and forward current gain are of particular use as output parameters. These functions can be defined as follows:

```
# MAXIMUM CUTOFF FREQUENCY
EXTRACT NAME="FT_MAX" MAX(G."COLLECTOR""BASE"/(6.28*C."BASE""BASE"))
#FORWARD CURRENT GAIN
EXTRACT NAME="PEAK GAIN" MAX(I."COLLECTOR"/I."BASE")
```

5.3   EXAMPLES OF SILVACO ATLAS SIMULATIONS

In this section we demonstrate several examples of the usage of the ATLAS simulation software for the purpose of making the user familiar with the syntax, to which one can always refer to by using the Silvaco Simulation Software Manuals. Furthermore, these examples serve to point the user to where and when analytical models fail and numerical device modeling is absolutely required. Examples of device structures simulated include: *pn*-diode, 25 nm MOSFET device, BJT, SOI, and MESFET device structures. Except for memory devices, in principle this selection of devices covers the most important problems encountered by engineers working either in industry or academia.

5.3.1   *pn*-Diode

Most modern diodes are based on *semiconductor pn-junctions*. In a *pn*-diode, conventional current can flow from the *p*-type side (the *anode*) to the *n*-type side (the *cathode*), but not in the opposite direction. Another type of semiconductor diode, the *Schottky diode*, is formed from the contact between a metal and a semiconductor rather than by a *pn*-junction. A semiconductor diode's current–voltage, or *IV*-characteristic curve is ascribed to the behavior of the so-called *depletion layer* or *depletion zone* which exists at the *pn-junction* between the differing semiconductors. When a *pn*-junction is first created, conduction band (mobile) electrons from the *n*-doped region diffuse into the *p*-doped region where there is a large population of holes with which the electrons recombine. When a mobile electron recombines with a hole, the hole vanishes and the electron is no longer mobile. The region around the *pn*-junction becomes depleted of *charge carriers* and thus behaves as an *insulator*. However, the *depletion width* cannot grow without limit. For each electron–hole pair that recombines, a positively-charged dopant ion is left behind in the *n*-doped region, and a negatively charged dopant ion is left behind in the *p*-doped region. As recombination proceeds and more ions are created, an increasing electric field develops through the depletion region which opposes the further flow of charge across the junction. At this point, there is a *built-in* potential across the depletion zone. If an external voltage is placed across the diode with the same polarity as the built-in potential, the depletion zone continues to act as an insulator preventing a significant electric current. However, if the polarity of the external voltage opposes the built-in potential, recombination can once again proceed resulting in substantial electric current through the *pn*-junction. For silicon diodes, the built-in potential is approximately 0.6 V. Thus, if an external current is forced through the diode, about 0.6 V will be developed across the diode such that the *p*-doped region is positive with respect to the *n*-doped region and the diode is said to be "turned on."

A diode's *IV*-characteristic (shown in Fig. 5.4) can be approximated by two regions of operation. Below a certain difference in potential between the two leads, the depletion layer has significant width, and the diode can be thought of as an open (nonconductive) circuit. As

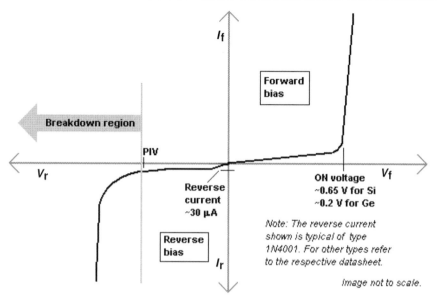

FIGURE 5.4: *I–V* characteristics of a *p–n* junction diode (not to scale)

the potential difference is increased, at some stage the diode will become conductive and allow charges to flow, at which point it can be thought of as a connection with zero (or at least very low) resistance. More precisely, the *transfer function* is *logarithmic*, but so sharp that it looks like a corner on a zoomed-out graph.

The *Shockley ideal diode equation* can be used to approximate the *pn*-diode's *IV*-characteristic

$$I = I_\text{s}\left[\exp\left(\frac{q V_\text{D}}{\eta k_\text{B} T}\right) - 1\right] \qquad (5.1)$$

where I is the diode current, I_S is a scale factor called the *saturation current*, q is the charge on an *electron* (the *elementary charge*), k_B is *Boltzmann's constant*, T is the absolute temperature of the *pn*-junction, and V_D is the voltage across the diode. The term $k_\text{B} T/q$ is the *thermal voltage*, sometimes written V_T, and is approximately 26 mV at room temperature. η (sometimes omitted) is the *emission coefficient*, which varies from about 1–2 depending on the fabrication process and semiconductor material. It is possible to use shorter notation. Putting $k_\text{B} T/q = V_\text{T}$ and $\eta = 1$, the relationship of the diode becomes:

$$I = I_\text{s}\left[\exp\left(\frac{V_\text{D}}{V_\text{T}}\right) - 1\right]. \qquad (5.2)$$

In a normal silicon diode at rated currents, the voltage drop across a conducting diode is approximately 0.6–0.7 V. The value is different for other diode types—Schottky diodes can be as low as 0.2 V and *light-emitting diodes* (LEDs) can be 1.4 V or more depending on the current.

Referring to the *IV*-characteristics in Fig. 5.4, in the reverse bias region for a normal *p–n* rectifier diode, the current through the device is very low (in the μA range) for all reverse voltages up to a point called the peak-inverse-voltage (PIV). Beyond this point a process called reverse breakdown occurs which causes the device to be damaged along with a large increase in current. For special purpose diodes like the avalanche or zener diodes, the concept of PIV is not applicable since they have a deliberate breakdown beyond a known reverse current such that the reverse voltage is "clamped" to a known value (called zener voltage). The devices however have a maximum limit to the current and power in the zener or avalanche region.

5.3.1.1   Diode Simulation Example

This is the simplest example that one can construct in ATLAS. It is a 1D problem that is treated as 2D for the purpose of setting up the structure file properly. For 2D simulations, the width of the device under consideration is by default equal to 1 μm. The purpose of this example is to demonstrate the following point: for moderately high doping densities of the diode, the depletion approximation is a good model and quite accurately estimates the peak electric field at the metallurgical junction. For the case of very high doping and/or asymmetric junctions the depletion approximation fails to accurately estimate the peak electric field, thus underestimating the voltage for which semiconductor breakdown occurs. For the purpose of examining this point, the user is asked to write an input deck that will allow him to calculate the charge density profile, the potential profile, and the electric field profile in equilibrium for a *pn*-diode with doping:

1. $N_A = N_D = 10^{16}$ cm^{-3}
2. $N_A = 10^{16}$ cm^{-3} and $N_D = 10^{18}$ cm^{-3}

For the two diode configurations the user is also asked to calculate the corresponding *IV*-characteristics. This will demonstrate to the user a very important point on how the diode turn-on voltage changes with increasing doping concentration on one side of the junction.

5.3.2   MOSFET Devices

The metal oxide semiconductor field-effect transistor (MOSFET) is by far the most common *field-effect transistor* in both *digital* and *analog* circuits. It was invented by Dawon Kahng and Martin Atalla at Bell Labs in 1960, and is composed of a metal (or polycrystalline silicon) gate separated from a semiconductor (either *p* or *n*-type) by an insulating gate material. Due to the fortuitously high quality of the oxide semiconductor interface in Si, most MOSFETs

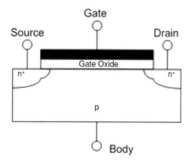

FIGURE 5.5: Cross-section of an NMOS

are fabricated on Si with SiO$_2$ as the insulator. Figure 5.5 shows a schematic cross-section of a MOSFET. Here *n*-type regions are diffused or implanted into a *p*-type substrate. With a positive gate-bias applied, minority carrier electrons are induced at the surface under the gate, creating a conducting *n*-type channel between the source and drain. Such a *n*-channel device is sometimes referred to as a NMOSFET. In contrast, a PMOSFET has the opposite structure (*p*-type source and drain regions, *n*-type substrate). Usually the semiconductor of choice is *silicon*, but some chip manufacturers, most notably *IBM*, have begun to use a mixture of *silicon* and *germanium* (SiGe) in MOSFET channels. Unfortunately, many semiconductors with better electrical properties than silicon, such as *gallium arsenide*, do not form good gate oxides and thus are not suitable for MOSFETs. IGFET is a related term meaning insulated-gate field-effect transistor, and is almost synonymous with "MOSFET," though it can refer to FETs with a gate insulator that is not oxide.

The gate terminal is a layer of *polysilicon* (polycrystalline silicon; why polysilicon is used will be explained below) placed over the channel, but separated from the channel by a thin layer of insulating silicon dioxide. When a voltage is applied between the gate and source terminals, the electric field generated penetrates through the oxide and creates a so-called "inversion channel" in the channel underneath. The inversion channel is of the same type—*p*-type or *n*-type—as the source and drain, so it provides a conduit through which current can pass. Varying the voltage between the gate and body modulates the *conductivity* of this layer and makes it possible to control the current flow between drain and source.

The operation of a MOSFET can be separated into three different modes, depending on the voltages at the terminals. For the NMOSFET the modes are:

1. *Cutoff or subthreshold mode:* When $V_{GS} < V_{th}$ where V_{th} is the *threshold voltage* of the device. The transistor is turned off, and there is no conduction between drain and source. While the current between drain and source should ideally be zero since the switch is turned off, there is a weak-inversion current or *subthreshold leakage*.

2. *Triode or linear region:* When $V_{GS} > V_{th}$ and $V_{DS} < V_{GS} - V_{th}$. The transistor is turned on, and a channel has been created which allows current to flow between the drain and source. The MOSFET operates like a resistor, controlled by the gate voltage. The current from drain to source for a long channel device is,

$$I_D = \frac{\mu_n C_{ox}}{2} \frac{W}{L} \left(2(V_{GS} - V_{th})V_{DS} - V_{DS}^2\right) \qquad (5.3)$$

where μ_n is the charge-carrier mobility, W is the gate width, L is the gate length, and C_{ox} is the capacitance at the gate.

3. *Saturation:* When $V_{GS} > V_{th}$ and $V_{DS} > V_{GS} - V_{th}$. The switch is turned on, and a channel has been created which allows current to flow between the drain and source. Since the drain voltage is higher than the gate voltage, a portion of the channel is turned off. The onset of this region is also known as pinch-off. The drain current is now relatively independent of the drain voltage (in a *first-order approximation*) and the current is only controlled by the gate voltage such that,

$$I_D = \frac{\mu_n C_{ox}}{2} \frac{W}{L}(V_{GS} - V_{th})^2 \qquad (5.4)$$

In *digital circuits* the transistors are only operated in the cutoff and triode modes. The saturation mode is mainly used in *analog circuit* applications.

The growth of digital technologies, like the *microprocessor*, has provided the motivation to advance MOSFET technology faster than any other type of silicon-based transistor. The principal reason for the success of the MOSFET was the development of digital *CMOS* logic, which uses *p*- and *n*-channel MOSFETs as building blocks. The great advantage of CMOS logic is that it allows no current to flow (ideally), and thus no *power* to be consumed, except when the inputs to *logic gates* are being switched. CMOS accomplishes this by complementing every nMOSFET with a pMOSFET and connecting both gates in such a way that whenever one is conducting, the other is not. This arrangement greatly reduces power consumption and heat generation. Overheating is a major concern in *integrated circuits*, since ever more transistors are packed into ever smaller chips. Another advantage of MOSFETs for digital switching is that the oxide layer between the gate and the channel prevents DC current from flowing through the gate, further reducing power consumption. The insulating oxide between the gate and channel effectively isolates a MOSFET in one logic state from earlier and consequent stages, which is vital because the gate of one MOSFET is usually driven by the output from a previous logic stage. This isolation makes it easier for designers to design logic stages independently. The MOSFET's strengths as the workhorse *transistor* in most *digital circuits* do not translate into supremacy in *analog circuits*. The *bipolar junction transistor* (BJT) has traditionally been the analog designer's transistor of choice, due largely to its high *transconductance* and unique properties. Nevertheless, MOSFETs are widely relied upon for analog purposes as well. Some analog circuits are designed solely using MOSFETs in a *fabrication* process specialized for digital circuits because it is advantageous to incorporate digital and analog circuits onto the same chip and digital fabrication processes are less expensive. Fabrication processes exist that incorporate BJTs and MOSFETs onto the same *die*, these mixed-transistor circuits are called *BiCMOS* (bipolar-CMOS) circuits. Ironically, the BJT has some advantages over the MOSFET in certain digital circuits; digital circuit designs can incorporate BJTs to speed signals in critical locations.

Over the past decades, the MOSFET has continually been scaled down in size; typical MOSFET channel lengths were once several *micrometres*, but modern *integrated circuits* are incorporating MOSFETs with channel lengths of about several tenths of nanometers. Until the late 1990s, this size reduction resulted in great improvement in the MOSFET operation with no deleterious consequences. Historically, the difficulties with decreasing the size of the MOSFET have been associated with the *semiconductor device fabrication* process. Smaller MOSFETs are desirable for three reasons. First, smaller MOSFETs allow more current to pass. Conceptually, MOSFETs are like *resistors* in the on-state, and shorter resistors have less resistance. Second, smaller MOSFETs have smaller gates, and thus lower gate *capacitance*. These first two factors contribute to lower switching times, and thus higher processing speeds. A third reason for MOSFET scaling is reduced area, leading to reduced cost. Smaller MOSFETs can be packed more densely, resulting in either smaller chips or chips with more computing power in the same area. Because the cost of fabricating a *semiconductor wafer* is relatively fixed, the cost of the individual *integrated circuits* is mainly related to the number of chips that can be produced per wafer. Hence, smaller IC's allow more chips per wafer, reducing the price per chip. Producing MOSFETs with channel lengths smaller than a micrometer is a challenge, and the difficulties of semiconductor device fabrication are always a limiting factor in advancing integrated circuit technology. Recently, however, the small size of the MOSFET has created operational problems.

5.3.2.1 Subthreshold Leakage

Because of small MOSFET geometries, the voltage that can be applied to the gate must be reduced to maintain reliability. To maintain performance, the *threshold voltage* of the MOSFET has to be reduced as well. As the threshold voltage is reduced, the transistor cannot be completely turned off, resulting in a weak-inversion layer which consumes power in the form of *subthreshold leakage* when the transistor should not be conducting. Subthreshold leakage, which was ignored in the past, now can consume upwards of half of the total power consumption of the chip.

5.3.2.2 Interconnect Capacitance

Traditionally, switching time was roughly proportional to the gate capacitance of gates. However, with transistors becoming smaller and more transistors being placed on the chip, *interconnect* capacitance (the capacitance of the wires connecting different parts of the chip) is becoming a large percentage of capacitance. Signals now have to travel through interconnects, which leads to increased delay and lower performance.

5.3.2.3 Heat Production

The ever-increasing density of MOSFETs on an integrated circuit is creating problems of substantial localized heat generation that can impair circuit operation. Circuits operate slower at high temperatures, and have reduced reliability and shorter lifetimes. Heat sinks and other cooling methods are now required for many integrated circuits including microprocessors. Power MOSFETs are at risk of *thermal runaway*. As their on-state resistance rises with temperature, the power loss on the junction rises correspondingly, generating further heat. When the heat sink is not able to keep the temperature low enough, the junction temperature may quickly and uncontrollably rise, resulting in failure of the device.

5.3.2.4 Gate Oxide Leakage

The gate oxide, which serves as the insulator between the gate and the channel, should be made as thin as possible to increase the channel conductivity and performance when the transistor is on, and to reduce subthreshold leakage when the transistor is off. However, with current gate oxides with a thickness of around 2 nm (which in silicon is 5 *atoms* thick) the phenomenon of *tunneling* leakage occurs between the gate and channel, leading to increased power consumption. Insulators (referred to as *high-k dielectrics*) that have a larger *dielectric constant* than *silicon dioxide*, such as group IVb metal silicates, e.g., *hafnium* and *zirconium* silicates and oxides, are now being researched to reduce the *gate leakage*. Increasing the dielectric constant of the gate oxide material allows a thicker layer while maintaining a high capacitance. The higher thickness reduces the tunneling current between the gate and the channel. An important consideration is the barrier height of the new gate oxide; the difference in *conduction band* energy between the semiconductor and the oxide (and the corresponding difference in *valence band* energy) will also affect the leakage current level. For the traditional gate oxide, silicon dioxide, the former barrier is approximately 3 eV. For many alternative dielectrics the value is significantly lower, somewhat negating the advantage of higher dielectric constant.

5.3.2.5 Process Variations

With MOSFETS becoming smaller, the number of atoms in the silicon that produce many of the transistor's properties is becoming fewer. During chip manufacturing, random process variation can affect the size of the transistor, which becomes a greater percentage of the overall transistor size as the transistor shrinks. The transistor characteristics become less deterministic, but more statistical. This statistical variation increases design difficulty.

The primary criterion for the gate material is that it is a good *conductor*. Highly-doped *polycrystalline silicon* is an acceptable, but certainly not ideal conductor, and it also suffers from some more technical deficiencies in its role as the standard gate material. There are a few reasons why polysilicon is preferable to a metal gate:

1. The threshold voltage (and consequently the drain to source on-current) is determined by the *work function* difference between the gate material and channel material. When metal was used as gate material, gate voltages were large (in the order of 3–5 V), the threshold voltage (resulting from the work-function difference between a metal gate and silicon channel) could still be overcome by the applied gate voltage (i.e., $|V_g - V_t| > 0$). As transistor sizes were scaled down, the applied signal voltages were also brought down (to avoid gate oxide breakdown, hot-electron reduction, power consumption reduction, etc.). A transistor with a high threshold voltage would become nonoperational under these new conditions. Thus, *polycrystalline silicon* (polysilicon) became the modern gate material because it is the same chemical composition as the silicon channel beneath the gate oxide. In inversion, the work-function difference is close to zero, making the threshold voltage lower and ensuring the transistor can be turned on.

2. In the MOSFET *IC fabrication* process, it is preferable to deposit the gate material prior to certain high-temperature steps in order to make better performing transistors. Unfortunately these high temperatures would melt metal gates, thus a high melting point material such as polycrystalline silicon is preferable to metal as a gate material. However, polysilicon is highly resistive (approximately 1000 times more resistive than metal) which reduces the signal propagation speed through the material. To lower the resistivity, dopants are added to the polysilicon. Sometimes additionally, high-temperature refractory metals such as *tungsten* are layered onto the top of the polysilicon (as a side effect of layering metal on the source and drain contacts) which decreases the resistivity. Such a blended material is called *z silicide*. The silicide–polysilicon combination has better electrical properties than polysilicon alone and still does not melt in subsequent processing. Also, the threshold voltage is not significantly higher than polysilicon alone, because the silicide material is not near the channel.

There are *depletion mode* MOSFET devices, which are less commonly used than the standard *enhancement mode* devices already described. These are MOSFET devices which are doped so that a channel exists even without any voltage applied to the gate. In order to control the channel, a negative voltage is applied to the gate, depleting the channel which reduces the current flow through the device. In essence, the depletion mode device is equivalent to a *normally closed* switch, while the enhancement mode device is equivalent to a *normally open* switch [1].

n-channel MOSFETs are smaller than p-channel MOSFETs and producing one type of MOSFET on a silicon substrate is cheaper and technically simpler. These were the driving

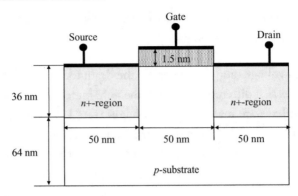

FIGURE 5.6: MOSFET device being simulated

principles in the design of *NMOS logic* which uses *n*-channel MOSFETs exclusively. However, unlike CMOS logic, NMOS logic consumes power even when no switching is taking place. With advances in technology, CMOS logic displaced NMOS logic in the 1980s to become the preferred process for digital chips.

5.3.2.6 MOSFET Simulation Example

The purpose of this simulation example is to introduce the user to modeling of the impact ionization process that leads to higher current densities and transistor breakdown at voltages smaller than the theoretically predicted ones. For this purpose, the user is asked to write a set of Silvaco ATLAS commands for modeling a MOSFET device structure, schematically shown in Figure 5.6. In the calculations it is required to use the appropriate model for low field mobility description in silicon inversion layers, velocity saturation effect, and the impact ionization model due to Selberherr. The oxide thickness of the simulated device is 1.5 nm, and the substrate doping is 10^{20} cm^{-3}. The junction depth is 0.36 nm and the total device depth, measured from the Si–SiO$_2$ interface, is 0.1 μm. For the doping of the source and drain regions, a donor concentration of 10^{20} cm^{-3} is to be used. The user is asked to vary the gate voltage from 0.8 to 1.4 V, in 0.2 V increments. For each gate voltage value, the user has to perform a drain voltage sweep from 0 to 2 V.

5.3.2.7 Listing of the Code Provided to the User

```
##############################################################
# This is the script for MOSFET simulation
##############################################################
#
mesh   space.mult=1.0
```

```
#
x.mesh loc=0.00 spac=0.005
x.mesh loc=0.05 spac=0.0002
x.mesh loc=0.075 spac=0.003
x.mesh loc=0.10 spac=0.0002
x.mesh loc=0.15 spac=0.005
#
y.mesh loc=-0.0015 spac=0.0005
y.mesh loc=0 spac=0.0004
y.mesh loc=0.036 spac=0.002
y.mesh loc=0.10 spac=0.005
# REGIONS AND ELECTRODES
region num=1 y.min=0 silicon
region num=2 y.max=0 oxide
elect  num=1  name=gate  x.min=0.05  length=0.05  y.min=-0.0015\
  y.max=-0.0015
elect num=2 name=source left length=0.04999 y.min=0.0 y.max=0.0
elect num=3 name=drain right length=0.05 y.min=0.0 y.max=0.0
elect num=4 name=substrate substrate

# DEVICE DOPING
doping uniform p.type conc=1.e19 y.min=0.036001
doping uniform  p.type  x.min=0.05  x.max=0.10  y.min=0  y.max=0.036\
  conc=1.e19
doping uniform n.type x.min=0.00 x.max=0.04999 y.min=0 y.max=0.036\
conc=1.e20
doping uniform n.type x.min=0.10001 x.max=0.15 y.min=0 y.max=0.036\
conc=1.e20

save outfile=mos_device_structure_0.str

go atlas
# IMPORT THE MESH
mesh inf=mos_device_structure_0.str master.in

# MATERIAL CONTACT INTERFACE AND MODELS
contact num=1 n.polysilicon
```

```
models print consrh cvt
# impact selb
# INITIAL SOLUTION
solve init
method newton trap
save outf=mos_init.str master.in

#############################################################
# RAMP THE GATE VOLTAGE FOR GENERATING THE FAMILY
# OF CURVES
#############################################################
solve prev
solve vgate=0.0 vstep=0.1 name=gate vfinal=0.1
method newton trap

solve vgate=0.2
method newton trap
save outf=mos_Vg_02.str master.in

solve vgate=0.4
method newton trap
save outf=mos_Vg_04.str master.in

solve vgate=0.6
method newton trap
save outf=mos_Vg_06.str master.in

solve vgate=0.8
method newton trap
save outf=mos_Vg_08.str master.in

solve vgate=1.0
method newton trap
save outf=mos_Vg_10.str master.in

solve vgate=1.2
method newton trap
save outf=mos_Vg_12.str master.in
```

```
solve vgate=1.4
method newton trap
save outf=mos_Vg_14.str master.in


###########################################################
# CALCULATE ID-VD CHARACTERISTIC FOR FIXED
# VALUE OF VG
###########################################################
load inf=mos_Vg_10.str master
log outf=mos_Vd_10.log
solve vdrain=0.0 vstep=0.1 vfinal=2.0 name=drain compliance=1.E-3\
cname=drain
method newton trap

load inf=mos_Vg_12.str master
log outf=mos_Vd_12.log
solve vdrain=0.0 vstep=0.1 vfinal=2.0 name=drain compliance=1.E-3\
cname=drain
method newton trap

load inf=mos_Vg_14.str master
log outf=mos_Vd_14.log
solve vdrain=0.0 vstep=0.1 vfinal=2.0 name=drain compliance=1.e-3\
cname=drain
method newton trap

# tonyplot -overlay mos_Vd_08.log mos_Vd_10.log \
# mos_Vd_12.log mos_Vd_14.log

Quit
```

5.3.2.8 Simulation Results

The simulated output characteristics for the MOSFET device structure from Figure 5.6 and substrate doping of 10^{20} cm^{-3} are shown in Figure 5.7. Notice the importance of the impact ionization process at high drain voltages, which results in an increase of current with increasing source-drain bias due to carrier multiplication.

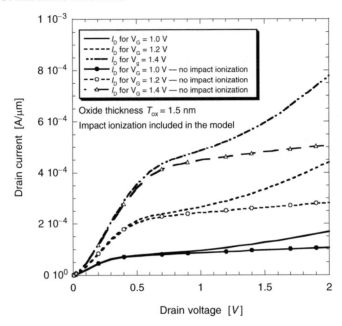

FIGURE 5.7: Output characteristic of the MOSFET device with substrate doping 10^{20} cm^{-3}

5.3.3   Simulation of BJT

A *bipolar junction transistor* (BJT) is another type of *transistor*. It is a three-terminal device and may work as an amplifying or switching device. Bipolar transistors are so named because the main conduction channel employs both *electrons* and *holes* to carry the main electric current. The bipolar junction transistor was invented in 1948 at the *Bell Telephone Laboratories* and enjoyed nearly three decades as the device of choice in the design of discrete and *integrated circuits*. Nowadays, the use of the BJT has declined in favor of the *MOSFET* and *CMOS* is now the technology of choice in the design of integrated circuits, as discussed in the previous section. Nevertheless, the BJT remains a major device that excels in some applications, such as the discrete circuit design, due to a very wide selection of BJT types available and also because of wide knowledge about the bipolar transistor characteristics. The BJT is also the choice for demanding analog circuits, both integrated and discrete. This is especially true in *very-high-frequency* applications, such as *radio-frequency* circuits for wireless systems. Bipolar transistors can be combined with MOSFET's to create innovative circuits that take advantage of the best characteristics of both types. This is called *BiMOS* and is increasing its areas of application.

A BJT consists of three differently *doped* semiconductor regions, the *emitter* region, the *base* region, and the *collector* region, comprised respectively of p type, n type and p type in a PNP transistor, and n type, p type, and n type in an NPN transistor. Each semiconductor region is connected to a terminal, appropriately labeled the *emitter* (E), *base* (B), and *collector*

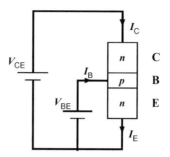

FIGURE 5.8: Structure and use of *npn* transistor

(C), as shown in Figure 5.8. The *base* is physically located between the *emitter* and the *collector* and is made from lightly doped, high-resistivity material. By varying the voltage across the base–emitter terminals very slightly, the current allowed to flow between the *emitter* and the *collector*, which are both heavily doped and hence low-resistivity regions, can be varied. This effect can be used to amplify the input current. BJTs can be thought of as voltage-controlled *current sources* but are usually characterized as *current amplifiers* due to the low impedance at the base. Early transistors were made from *germanium* but most modern BJTs are made from *silicon*.

An *npn* bipolar transistor can be considered as two *diodes* connected *anode* to anode. In normal operation, the emitter–base *junction* is forward biased and the base–collector junction is reverse biased. In an *npn*-type transistor for example, electrons from the emitter are injected into the base by the forward biased emitter-base n-p junction. These electrons in the base are then *minority* carriers, and there are plenty of *holes* with which to recombine. The base is always made very thin so that most of the electrons diffuse over to the collector before they recombine with holes. The collector–base junction is reverse biased to prevent the flow of holes, while electrons are swept into the collector by the electric field around the junction. The proportion of electrons able to penetrate the base and reach the collector is approximately constant in most cases. However, the heavy doping (low resistivity) of the emitter region and light doping (high resistivity) of the base region mean that many more electrons are injected into the base, and therefore reach the collector, than there are holes injected into the emitter. The base current is the sum of the holes injected into the emitter and the electrons that recombine in the base— both small proportions of the total current. Hence, a small change of the base current can translate to a large change in electron flow between emitter and collector. The ratio of these currents I_c/I_b, called the *current gain*, and represented by β or h_{fe}, is typically 100 or more. It is important to keep the base region as thin and as free from defects as possible, in order to minimize recombination losses of the *minority carriers*.

Figures 5.8 is a schematic representation of an *npn*-transistor connected to two voltage sources. To make the transistor conduct appreciable current (on the order of 1 mA) from C to

E, V_{BE} must be equal to or slightly greater than the *cut-in* voltage. The cut-in voltage is usually between 600 mV and 700 mV for silicon-based BJTs. This applied voltage causes the lower *pn*-junction to "turn-on" allowing a flow of electrons from the emitter into the base. Because of the electric field existing between base and collector (caused by V_{CE}), the majority of these electrons cross the upper *pn*-junction into the collector to form the collector current, I_C. The remainder of the electrons exit the base connection to form the base current, I_B. As shown in the diagram, the emitter current, I_E, is the total transistor current which is the sum of the other terminal currents. That is:

$$I_E = I_B + I_C$$

(Note: In this diagram, the arrows representing current point in the direction of the electric or *conventional current*—the flow of electrons is in the opposite direction of the arrows since electrons carry negative *electric charge*). The ratio of this collector current to this base current is called the DC current gain. This gain is usually quite large and is often 100 or more. It should also be noted that the base current is related to V_{BE} exponentially. For a typical transistor, increasing V_{BE} by just 60 mV increases the base current by a factor of 10!

Transistors have different regions of operation. In the "linear" region, collector–emitter current is approximately *proportional* to the base current but many times larger, making this the ideal mode of operation for current *amplification*. The BJT enters "saturation" when the base current is increased to a point where the external circuitry prevents the collector current from growing any larger. At this point, the C–B junction also becomes forward biased. A residual voltage drop of approximately 100–300 mV (depending on the amount of base current) then remains between collector and emitter. Less commonly, bipolar transistors are operated with emitter and collector reversed, thus a base–collector current can control the emitter–collector current. The current gain in this mode is much smaller (i.e., 2 instead of 100), and it is not a value that is controlled by manufacturers, so it can vary dramatically among transistors. A transistor is said to operate in the "cutoff" region when the base–emitter voltage is too small for any significant current to flow. In typical BJTs manufactured from silicon, this is the case below 0.7 V or so. BJTs that operate only in "cut off" and "saturation" regions can by viewed as electronic *switches*.

Because of its temperature sensitivity, the BJT can be used to measure temperature. Its nonlinear characteristics can also be used to compute *logarithms*. The *germanium* transistor was more common in the 1950s and 1960s, and while it exhibits a lower "cut off" voltage, making it more suitable for some applications, it also has a greater tendency to exhibit thermal runaway. The *Heterojunction Bipolar Transistor* (*HBT*) is an improvement of the BJT that can handle signals of very high frequencies up to several hundred GHz. It is common nowadays in ultrafast circuits, mostly RF systems. Exposure of the transistor to *ionizing radiation* causes *radiation damage.* Radiation causes a buildup of "defects" in the base region that act as *recombination centers.* This causes gradual loss of gain of the transistor.

5.3.3.1 BJT Simulation Example

The simulation of a BJT is a little bit more complex when compared to the simulation of a MOSFET presented in the previous section. The reason for the more problematic convergence of the BJT transistor is that the base contact is a current-controlled current contact. Therefore, to achieve convergence, the base contact is initially defined as a voltage-controlled contact and the applied voltage on it is swept slowly from 0 to the turn-on voltage of the diode. Once steady-state condition is established the base contact is redefined as a current-controlled current contact. The aim of the exercise presented below is to use physically based ATLAS simulator to study basic characteristics of bipolar junction transistor (BJT) and to extract the small-signal *h*-parameters from the family of DC characteristics being simulated.

5.3.3.2 Exercise Description

a) Start with the attached input deck, add few more additional statements, and obtain the complete DC *IV*-characteristics of the Si BJT in the common-emitter configuration. In the output characteristics, use

$$I_B = 1E - 6A \text{ to } 7E - 6 \text{ A},$$

with increments of 1E-6 A.

b) Write the appropriate extract statements to calculate the common-emitter amplification factor β ($\Delta I_C/\Delta I_B$) for $V_{CE} = 2.0$ and 4.0 V as a function of the collector current. Why do you observe decrease in β at large current densities? Use physical reasoning for explaining the observed trend in the common-emitter amplification factor.

c) From the output characteristics extract the value of the Early voltage.

d) Using AC simulations, obtain the small-signal *h*-parameters (h_{ie}, h_{re}, h_{fe}, and h_{oe}) of the Si BJT for $V_{CE} = 3.0$ V as a function of the collector current I_C. Use frequency of 10 kHz, for which the *h*-parameter model is appropriate and no parasitic capacitance effects are significant.

5.3.3.3 Listing of the Basic Code

```
go atlas
TITLE structure of a bipolar junction transistor
#
Mesh
x.m l=0  spacing=0.15
```

```
x.m l=0.8 spacing=0.15
x.m l=1.0  spacing=0.03
x.m l=1.5 spacing=0.12
x.m l=2.0 spacing=0.15
#
y.m l=0.0 spacing=0.006
y.m l=0.04 spacing=0.006
y.m l=0.06 spacing=0.005
y.m l=0.15 spacing=0.02
y.m l=0.30 spacing=0.02
y.m l=1.0 spacing=0.12
#
region num=1 silicon
#
electrode num=1 name=emitter left length=0.8
electrode num=2 name=base right length=0.5 y.max=0
electrode num=3 name=collector bottom
#
doping reg=1 uniform n.type conc=5e15
doping reg=1 gauss n.type conc=1e18 peak=1.0 char=0.2
doping reg=1 gauss p.type conc=1e18 peak=0.05 junct=0.15
doping reg=1 gauss n.type conc=5e19 peak=0.0 junct=0.05 x.right=0.8
doping reg=1 gauss p.type conc=5e19 peak=0.0 char=0.08 x.left=1.5
#
# Set bipolar models
models conmob fldmob consrh auger print numcarr=2
contact name=emitter n.poly surf.rec
#
# Gummel plot
method  newton autonr trap
solve init
solve vcollector=0.025
solve vcollector=0.1
solve vcollector=0.25 vstep=0.25 vfinal=2 name=collector
#
solve vbase=0.025
solve vbase=0.1
```

```
solve vbase=0.2
#
log outf=bjt_0.log
solve vbase=0.3 vstep=0.05 vfinal=1 name=base
tonyplot bjt_0.log
#
# IC/VCE with constant IB
#
# Ramp Vb
#
log off
solve init
solve vbase=0.025
solve vbase=0.05
#
solve vbase=0.1 vstep=0.1 vfinal=0.7 name=base
#
# Switch to current boundary conditions
contact name=base current
#
# Ramp IB and save solutions
solve ibase=1.e-6
save outf=bjt_1.str master
#
solve ibase=2.e-6
save outf=bjt_2.str master
#
solve ibase=3.e-6
save outf=bjt_3.str master
#
solve ibase=4.e-6
save outf=bjt_4.str master
#
solve ibase=5.e-6
save outf=bjt_5.str master
#
log outf=bjt_iv.log
```

```
#
# Load in each initial guess file and ramp VCE
load inf=bjt_1.str master
log outf=bjt_1.log
solve vcollector=0.0 vstep=0.25 vfinal=5.0 name=collector
#
load inf=bjt_2.str master
log outf=bjt_2.log
solve vcollector=0.0 vstep=0.25 vfinal=5.0 name=collector
#
load inf=bjt_3.str master
log outf=bjt_3.log
solve vcollector=0.0 vstep=0.25 vfinal=3.0 name=collector
save outf=data_for_ac.str master
solve vcollector=3.25 vstep=0.25 vfinal=5.0 name=collector
#
load inf=bjt_4.str master
log outf=bjt_4.log
solve vcollector=0.0 vstep=0.25 vfinal=5.0 name=collector
#
load inf=bjt_5.str master
log outf=bjt_5.log
solve vcollector=0.0 vstep=0.25 vfinal=5.0 name=collector
#
quit
```

5.3.3.4 Simulation Results

The simulation results that show the finite element mesh set-up and its refining at the device active region are shown on the left panel of Figure 5.9. The corresponding potential profile in the device for zero applied bias is shown on the panel on the right of Figure 5.9.

One of the most important characteristics graphs that describe the operation of a BJT as an amplifier is the Gummel plot (Figure 5.10), which gives the variation of the base and collector current with applied base bias, thus providing one directly the current amplification of the transistor. The higher the ratio between the collector and the base current, for a larger range of emitter–base voltages, the better the amplification characteristics of the transistor.

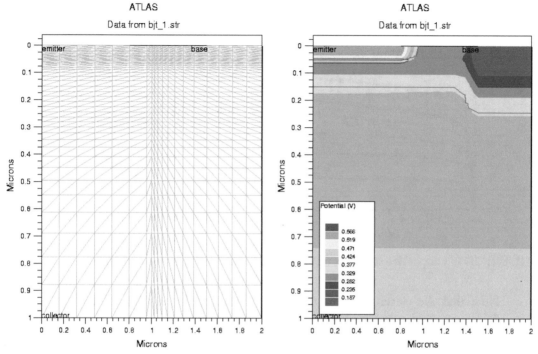

FIGURE 5.9: Mesh set-up (*left panel*) and potential profile (*right panel*)

The BJT output characteristics and the extraction of the Early voltage, obtained with the Silvaco ATLAS simulation software, are shown in the left and the right panel of Figure 5.11. It is clear that the Early voltage for this BJT is around 30 V, which is a rather high value.

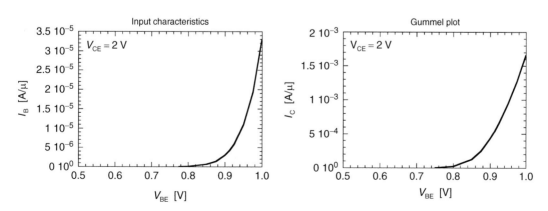

FIGURE 5.10: Gummel plot

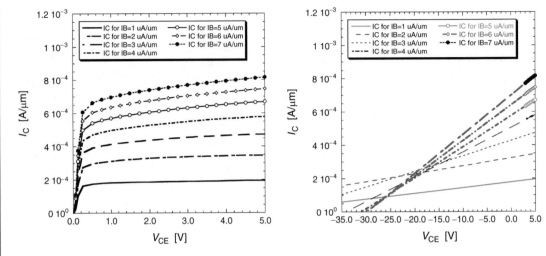

FIGURE 5.11: Output characteristics (*left panel*) and extraction of the Early voltage

Note that the current amplification factor varies with the collector current, i.e., the applied collector–emitter voltage due to several factors, one of them being the base-width modulation. The variation of the current amplification with the collector current is shown in Figure 5.12.

At low frequencies, the small-signal representation of a BJT in the common-emitter configuration is the following:

The *h*—(hybrid) parameters that appear in the circuit shown in Figure 5.13 are found from the relationship of the input and output voltages and currents in this two-port network:

$$v_{be} = h_{ie}i_b + h_{re}v_{ce},$$
$$i_c = h_{fe}i_b + h_{oe}v_{ce}.$$

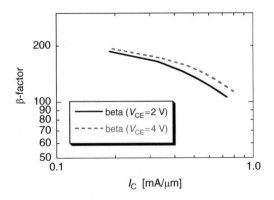

FIGURE 5.12: Amplification factor versus collector current

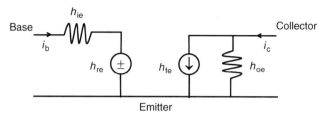

FIGURE 5.13: Two-port description of the transistor with h-parameters

This description leads to the following definition of the h-parameters:

$$h_{ie} = \left.\frac{v_{be}}{i_b}\right|_{v_{ce}=0} = \left.\frac{\partial V_{BE}}{\partial I_B}\right|_{V_{CE}=\text{const.}} \quad , h_{fe} = \left.\frac{i_c}{i_b}\right|_{v_{ce}=0} = \left.\frac{\partial I_C}{\partial I_B}\right|_{V_{CE}=\text{const.}}$$

$$h_{re} = \left.\frac{v_{be}}{v_{ce}}\right|_{i_b=0} = \left.\frac{\partial V_{BE}}{\partial V_{CE}}\right|_{I_B=\text{const.}} \quad , \quad h_{oe} = \left.\frac{i_c}{v_{ce}}\right|_{i_b=0} = \left.\frac{\partial I_C}{\partial V_{CE}}\right|_{I_B=\text{const.}}$$

One can use the above definitions of the h-parameters to generate the proper DC curves from which one can extract all four small-signal parameters for $V_{CE} = 3$ V. The h-parameters presented on these graphs were calculated in the following manner:

(1) For $V_{CE} = 3$ V, a sweep in the base current I_B was done and the parameters h_{ie} and h_{fe} were calculated using:

$$h_{ie} = \left.\frac{\partial V_{BE}}{\partial I_B}\right|_{V_{CE}=\text{const.}} \quad \text{and} \quad h_{fe} = \left.\frac{\partial I_C}{\partial I_B}\right|_{V_{CE}=\text{const.}}$$

(2) For given I_B, a sweep in the collector current was made and the h_{re} and h_{oe} parameters were calculated using:

$$h_{re} = \left.\frac{\partial V_{BE}}{\partial V_{CE}}\right|_{I_B=\text{const.}} \quad \text{and} \quad h_{oe} = \left.\frac{\partial I_C}{\partial V_{CE}}\right|_{I_B=\text{const.}}$$

The normalized h-parameters in the common-emitter configuration are shown in Figures 5.14 and 5.15.

5.3.4   Simulation of SOI Devices

The aim of this section is to give the reader more insight into the operation of SOI devices, and the different trends observed in the device output and subthreshold transfer characteristics for fully-depleted (FD) and partially-depleted (PD) SOI devices. Figure 5.16 shows the basic configuration of n-channel and accumulation mode p-channel SOI devices

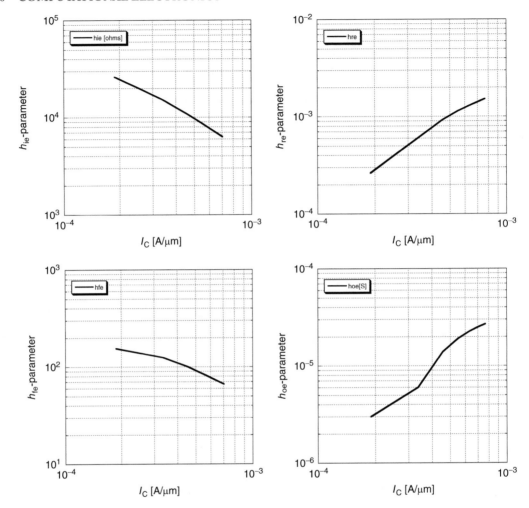

FIGURE 5.14: h-parameters variation with the collector current

For the n-channel devices, there are three modes of operation

- Thick-film (Partially-depleted) PD-SOI devices: $x_{dmax} < t_{Si}$
- Thin-film (fully-depleted) FD-SOI devices: $x_{dmax} > t_{Si}$
- Medium film SOI devices: $x_{dmax} < t_{Si} < 2x_{dmax}$

Figure 5.17 illustrates the differences between FD and PD-SOI devices. Thick-film SOI devices exhibit the following properties:

- No interaction between front and back depletion zones
- When the body (neutral region) is grounded, the PD-SOI device behaves as a bulk device

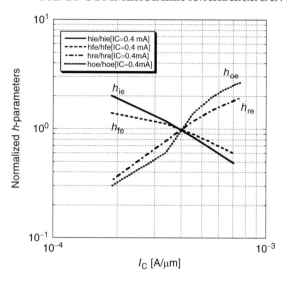

FIGURE 5.15: Normalized h-parameters in common-emitter configuration

FIGURE 5.16: n-channel and p-channel SOI devices

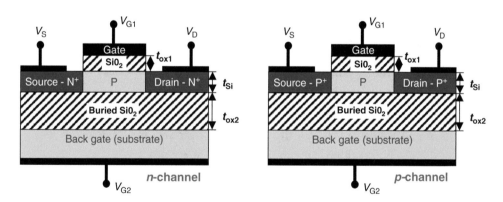

FIGURE 5.17: Depletion region of PD and FD devices

- When the body is left floating, two parasitic effects can occur:
  1. Kink effect
  2. Presence of a parasitic open-base *npn* BJT between source and drain

Thin-film SOI devices, on the other hand, have the following properties:

- Low electric fields
- High transconductance
- Excellent short-channel behavior
- Quasi-ideal subthreshold slope

The effects described below require special care when modeling SOI devices:

1. Impact ionization process and high fields
   a) Kink effect in PD-SOI *n*-channel devices. The effect is not present in FD SOI devices because of two reasons:
      i) Lower electric fields near the drain region
      ii) Holes recombine at the source without having to raise the body potential
   b) Hot-electron degradation—more pronounced in SOI devices when compared to their bulk counterparts
      i) Self-heating effects: SOI devices are thermally isolated from the substrate by the buried insulator. Hence, the removal of heat within the device is less efficient than in bulk, which can elevate device temperature.
      ii) Quantum Effects: For extremely thin SOI films, there will be associated increase in the energy minimum of the conduction band due to the quantum-mechanical size-quantization effect.
2. The Silvaco ATLAS recommendations on models include:
   a) Mobility: KLA + FLDMOB
   b) Interface charge: INTERFACE statement used for both interfaces (front and back)
   c) Recombination: SRH
   d) Band-gap narrowing: BGN
   e) Carrier generation: IMPACT and AUGER
   f) Lattice heating: LAT.TEMP on MODELS statement
   g) Carrier heating: Switch additional balance equations model

Silvaco ATLAS recommendations on numerical method chosen are as follows. Since problems can occur when having the floating body, because of the poor initial guess for the potential on it, the choice METHOD GUMMEL NEWTON initially performs a Gummel iteration to obtain improved initial guess for the Newton solution.

5.3.4.1 Exercise Description

An input deck for simulation of a particular device structure is users' starting point for simulation of the I_D–V_D characteristics of a partially-depleted SOI device for three different values of the gate voltage. To properly account for the so-called "kink" effect, Selberherr's impact ionization model is included.

a) The user is asked to run this example and plot the output characteristics of the PD-SOI device. To examine the role of the impact ionization process, first, exclude the impact ionization model and repeat the previous simulation runs. Discuss the differences in the two sets of output characteristics for the cases when you have included and excluded the impact ionization process in your model.

b) Next, the user is asked to modify the input deck, so that it only calculates the device transfer characteristics for $V_D = 0.1$ V and V_G ramped to 1.5 V in 0.1 V increments. Next, he/she needs to vary the thickness of the silicon layer from 0.3 μm (this is the thickness of the Si film specified in the example) down to 0.1 μm, with 50 nm decrements in thickness. In this process, the user has to make sure that the thickness of the underlying buried oxide layer remains the same (it equals 0.4 μm). For each of these SOI device structures, he/she has to plot the transfer characteristics and extract the magnitude of the subthreshold slope. Finally, a plot of the subthreshold slope as a function of the Si film thickness is required accompanied by a discussion of the body thickness value at which the device transitions from partially-depleted to fully-depleted SOI device structure.

5.3.4.2 Listing of the Code Provided to the User

```
go atlas
TITLE SOI device simulation
#
# 0.2um of silicon on 0.4um oxide substrate
#
mesh    space.mult=1.0
#
x.mesh loc=0.00    spac=0.50
x.mesh loc=1.15    spac=0.02
```

```
x.mesh loc=1.5    spac=0.1
x.mesh loc=1.85   spac=0.02
x.mesh loc=3      spac=0.5
#
y.mesh loc=-0.017 spac=0.02
y.mesh loc=0.00   spac=0.005
y.mesh loc=0.1    spac=0.02
y.mesh loc=0.2    spac=0.01
y.mesh loc=0.6    spac=0.25
#
region      num=1 y.max=0 oxide
region      num=2 y.min=0 y.max=0.2 silicon
region      num=3 y.min=0.2 oxide
#
#*********** define the electrodes ************
# #1-GATE #2-SOURCE #3-DRAIN #4-SUBSTRATE(below oxide)
#
electrode      name=gate    x.min=1 x.max=2 y.min=-0.017 y.max=-0.017
electrode      name=source x.max=0.5 y.min=0 y.max=0
electrode      name=drain x.min=2.5 y.min=0 y.max=0
electrode      substrate
#
#*********** define the doping concentrations *****
#
doping     uniform conc=2e17 p.type reg=2
doping     gauss n.type conc=1e20 char=0.2 lat.char=0.05 reg=2 x.r=1.0
doping     gauss n.type conc=1e20 char=0.2 lat.char=0.05 reg=2 x.l=2.0
save       outf=soiex01_0.str
tonyplot   soiex01_0.str -set soiex01_0.set
#
# set interface charge separately on front and back oxide interfaces
interf     qf=3e10 y.max=0.1
interf     qf=1e11 y.min=0.1
#
# set workfunction of gate
contact      name=gate n.poly
#
```

```
# select models
models        conmob srh auger bgn fldmob print
#
solve init
#
# do IDVG characteristic
#
method        newton    trap
solve         prev
solve         vgate=-0.2
solve         vdrain=0.05
solve         vdrain=0.1
#
# ramp gate voltage
log           outf=soiex01_1.log master
solve         vgate=0.1 vstep=0.1 name=gate vfinal=1.5
#
# plot resultant IDVG threshold voltage curve
tonyplot      soiex01_1.log -set soiex01_1.set
#
# plot resultant IDVG subthreshold slope curve
tonyplot      soiex01_1.log -set soiex01_2.set
#
#
extract name="subvt" \
        1.0/slope(maxslope(curve(v."gate",log10(abs(i."drain")))))
#
#
Extract name="vt" (xintercept(maxslope(curve(v."gate", \
  abs(i."drain")))) \
        - abs(ave(v."drain"))/2.0)
#
Quit
```

5.3.4.3 Simulation Results

The output characteristics of the partially depleted (left panel) and fully-depleted (right panel) SOI device structure are shown in Figure 5.18. Note the existence of the Kink-effect in the

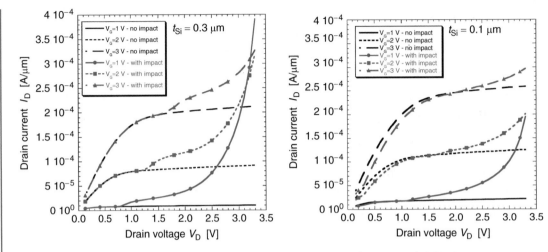

FIGURE 5.18: Output characteristics of PD (*left panel*) and FD (*right panel*) SOI devices

partially depleted device due to impact ionization holes modifying the body potential. The kink-effect has completely disappeared in the fully-depleted device because the generated holes at high drain bias (because of the impact ionization process) are swept very fast by the source contact and do not lead to modification of the body potential.

The variation of the subthreshold slope (left panel) and the device subthreshold characteristics (right panel) for different Si film thickness are shown in Figure 5.19. Note that the FD device has much better turn-off characteristics when compared to the PD device. For this

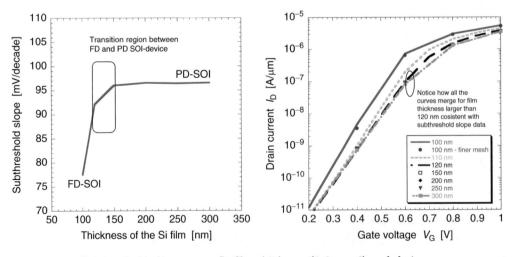

FIGURE 5.19: Subthreshold slope versus Si film thickness (*left panel*) and drain current versus gate voltage (*right panel*)

particular device geometry, the transition from PD to FD behavior occurs around 100 nm Si film thickness.

5.3.5   Gate Tunneling Models

For submicrometer devices, due to the smaller oxide thickness, there is significant conductance being measured on the gate contact. The finite gate current gives rise to the following effects:

- Negative—degradation in the device operating characteristics with time due to oxide charging, larger off-state power dissipation, etc.
- Positive—Utilized in nonvolatile memories where the gate current is used to program and erase charge on the "floating contact." Examples include FLASH, FLOTOX, EEPROM.

There are two different types of conduction mechanisms to the insulator layer included in the Silvaco ATLAS:

- Tunneling: Fowler–Nordheim or direct tunneling process
- Hot-carrier injection: Lucky electron or Concannon model

Regarding the Lucky electron model, an electron is emitted into the oxide when it gains sufficient energy to overcome the insulator/semiconductor barrier. The Concannon model is similar to the Lucky electron model, but assumes a non-Maxwellian high energy tail of the distribution function. This description requires the solution of the energy balance equation for carrier temperature. In the process of oxide charging, electrons at the drain end of the channel of a MOSFET device have sufficient energy to overcome the barrier at the Si/SiO_2 interface and be trapped at the oxide. Since the effect is cumulative, it limits the useful "life" of the device. To reduce oxide charging, LDD regions are usually used. The various oxide charging mechanisms, that lead to threshold voltage shift, are summarized in Figure 5.20.

The three types of tunneling processes occurring in MOS capacitors are schematically shown in Figure 5.21. For oxide thickness greater than 4 nm, Fowler–Nordheim tunneling process dominates. For oxide thicknesses less than 4 nm direct tunneling becomes important. Note that for a given electric field the current due to Fowler–Nordheim tunneling process is independent of the oxide thickness. On the other hand, the direct tunneling current is strongly correlated with the oxide thickness. The Fowler–Nordheim tunneling process is usually described using the WKB approximation, details of which can be found in many textbooks. In Silvaco ATLAS, the Fowler–Nordheim tunneling currents are calculated using the following expressions:

$$J_{FN} = \mathbf{F.AE} \cdot E^2 \exp\left(-\mathbf{F.BE}/E\right),$$
$$J_{FP} = \mathbf{F.AH} \cdot E^2 \exp\left(-\mathbf{F.BH}/E\right),$$

$$(5.5)$$

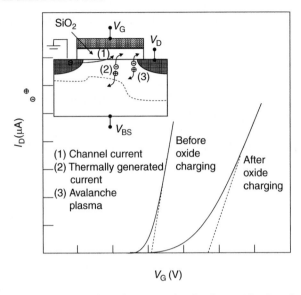

FIGURE 5.20: Schematic description of mechanisms that lead to oxide charging

where E is the magnitude of the electric field in the oxide, **F.AE, F.BE, F.AH,** and **F.BH** are model parameters that can be defined via the MODEL statement.

Note that in state-of-the-art devices, as the oxide thickness decreases, the gate current increases. In ultra-small devices, it will eventually dominate the off-state leakage current (drain current for $V_G = 0$ V). The gate current as a function of technology generation is shown in the following figure. The results shown in Figure 5.22 suggest that for channel lengths less than 50 nm it is critical to take into account gate leakage current when estimating the device power dissipation.

A schematic description of the Lucky electron model, also implemented in ATLAS simulation software, is shown in Figure 5.23. P_1 is the probability that the electron gains sufficient energy from the electric field to overcome the potential barrier. P_2 is the probability for

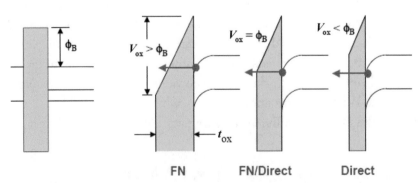

FIGURE 5.21: Description of the three most common tunneling mechanisms in a MOSFET device

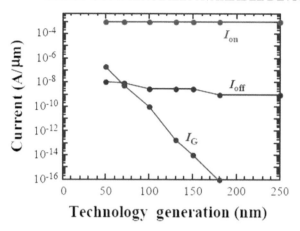

FIGURE 5.22: Gate current as a function of technology generation

redirecting collisions to occur, to send the electron toward the semiconductor/oxide interface. The probability that the electron will travel toward the interface without loosing energy is denoted by P_3. Finally, P_4 is the probability that the electron will not scatter in the image potential.

The various probabilities that appear in the description of the Lucky electron model are calculated using:

$$P_1 = \frac{1}{\lambda E_x} \exp\left(-\frac{E}{\lambda E_x}\right) dE$$

$$P_2 = \frac{1}{2\lambda_r}\left(1 - \sqrt{\frac{\Phi_B}{E}}\right)$$

$$P_3 = \exp\left(-\frac{y}{\lambda_r}\right)$$

$$P_4 = \exp\left(-\frac{x_0}{\lambda_{ox}}\right)$$

(5.5)

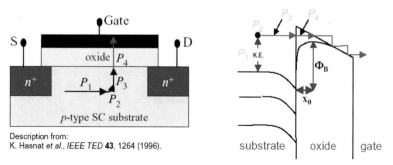

Description from:
K. Hasnat *et al.*, *IEEE TED* **43**, 1264 (1996).

FIGURE 5.23: Schematic description of the Lucky electron model

where E is the energy, λ is the scattering mean free path, λ_r is the redirection mean-free path, λ_{ox} is the mean-free path in the oxide, and the barrier height Φ_B at the semiconductor/oxide interface is calculated using

$$\Phi_B = \Phi_{B0} - \alpha E_{ox}^{1/2} - \beta E_{ox}^{2/3}, \tag{5.6}$$

where the first term on the RHS represents the zero-field barrier height, the second one describes the barrier lowering due to the image potential, and the third term accounts for probability for tunneling. In the Silvaco ATLAS implementation of the Lucky electron model, the probabilities P_1 and P_2 have actually been merged together. It is activated via the MODEL statement by the parameters HEI (hot- electron injection) or HHI (hot-hole injection).

5.3.6   Simulation of a MESFET

MESFET stands for Metal-Semiconductor Field Effect Transistor. It is quite similar to a Junction Field Effect Transistor (*JFET*) in construction and terminology. The difference is that instead of using a *pn*-junction for a gate, a *Schottky* (metal–semiconductor) junction is used. MESFETs are usually fabricated in GaAs, InP, or SiC, and hence are faster but more expensive than silicon-based JFETs or *MOSFETs*. MESFETs are operated up to approximately 30 GHz and are commonly used for *microwave* frequency *communications* and *radar*. From a digital *circuit design* perspective, it is difficult to use MESFETs as the basis for large-scale digital *integrated circuits*. The cross section of a simple GaAs MESFET is shown in Figure 5.24.

The *n*-type conducting channel in GaAs is produced by implanting the substrate with Si donors. Two metallization processes are used during the fabrication for the ohmic for source and drain contacts, and for the Schottky gate contact. The application of negative bias reverse

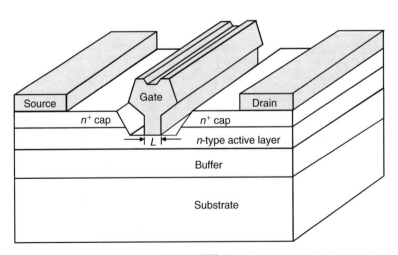

FIGURE 5.24: Schematics description of a MESFET device

biases the gate channel junction and increases the thickness of the depletion region. This affects the reverse bias drain-source current since the width of the conducting channel is controlled by the gate voltage. Usually the Schottky gate contact is reverse biased to reduce the gate leakage current. The device can be either a depletion mode or an enhancement mode device depending upon the whether the channel is pinched-off or not for $V_G = 0$. Since the Schottky barrier height is an important parameter in modeling MESFET devices, and since this parameter cannot be theoretically predicted with any accuracy, the only way to obtain an accurate value for the Schottky barrier height is via measurements. The determination of the Schottky barrier height is usually done using internal photoemission experiment. In this setup the semiconductor is illuminated with monochromatic light and the photocurrent is measured as a function of the wavelength, as shown in Figure 5.25. In Table 5.1 we summarize measured Schottky barrier heights on n-type silicon and GaAs.

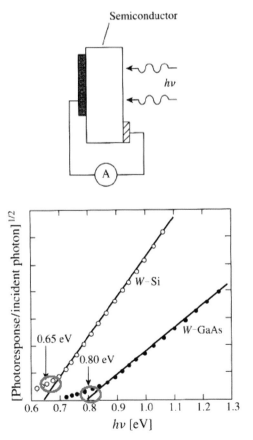

FIGURE 5.25: *Top panel*—internal photoemission setup. *Bottom panel*—normalized photoresponse versus wavelength

TABLE 5.1:		
	SCHOTTKY BARRIER HEIGHT (V)	
METAL	Fi_bn [Si]	Fi_bn {GaAs}
Al	0.72	0.80
Au	0.80	0.90
Pt	0.90	0.84
W	0.67	0.80

5.3.6.1 MESFET Simulation Example

The purpose of this example is to demonstrate to the user how one can calculate the ac S-parameters for a simple MESFET structure. It illustrates (1) the MESFET structure definition using ATLAS syntax; (2) Setting of the GaAs mobility models and gate workfunctions; (3) the I_d/V_{ds} characteristics for $V_{gs} = 0$; (4) the AC analysis at a single DC bias at different frequencies; (5) Conversion of capacitance and conductance data from ATLAS into S-parameters; and (6) the Smith chart of MESFET—S-parameters representation using TonyPlot. The geometry and the doping profile of the MESFET device are described using the ATLAS structural syntax. The ELECTRODE statements specify the names and positions of the electrodes. The workfunction of the gate contact is set using CONTACT. GaAs mobility models for concentration and field-dependence are set in the MODELS statement. The numerical methods used at the initial stage of the simulation are conservative. The statement METHOD GUMMEL NEWTON specifies that the decoupled (Gummel) method is used at the beginning of the simulation for each bias point. This switches to the coupled (Newton) method if convergence is not obtained. This is the most robust method for the initial bias steps and for complex devices, as discussed earlier in the Chapter. It does consume more CPU time and can lead to problems at high current levels. If these problems occur, the statement METHOD NEWTON can be used. The DC simulation proceeds by ramping the drain voltage in the SOLVE statement until $V_{ds} = 3.0$ V. After this a log file is opened and AC analysis begins. The full sweep of frequencies is performed on the line:

```
SOLVE AC.ANALYSIS DIRECT FREQUENCY=1.E9 FSTEP=2.E9 NFSTEPS=20
```

The AC analysis in ATLAS calculates the real and imaginary current components from a small AC signal on top of the existing DC solution. From these currents ATLAS calculates the conductance and capacitance between each pair of electrodes. The S-parameters are calculated by setting the S.PARAM on the log statement preceding the AC analysis. *S*-parameter analysis assumes the device is a two-port device with four terminals. The INPORT and OUTPORT are used to assign the respective electrodes.

5.3.6.2 Listing of the Code Provided to the User

```
go atlas
Title MBE Epitaxial GaAs MESFET - S parameters calculation

# Define the mesh

mesh  space.mult=1.0
#
x.mesh loc=0.00 spac=0.3
x.mesh loc=2.3 spac=0.02
x.mesh loc=2.7 spac=0.02
x.mesh loc=5 spac=0.3
#
y.mesh loc=0.00 spac=0.01
y.mesh loc=0.04 spac=0.03
y.mesh loc=0.12 spac=0.02
y.mesh loc=6 spac=1.0

# Region specification: Defined as two regions for different
  properties

region     num=1 GaAs x.min=0 x.max=5 y.min=0 y.max=0.12
region     num=2 GaAs x.min=0 x.max=5 y.min=0.12 y.max=6.12
# Electrode specification

elec       num=1  name=source  x.min=0.0 y.min=0.0 x.max=1.0 y.max=0.0
elec       num=2  name=drain   x.min=4.0 y.min=0.0 x.max=5.0 y.max=0.0
elec       num=3  name=gate    x.min=2.35 length=0.3

# Doping specification

doping region=1 uniform conc=1.0e17 n.type
doping region=2 uniform conc=1.0e15 p.type

# Set models, material and contact parameters
```

```
contact num=3 work=4.77
models region=1 print conmob fldmob srh optr
models region=2 srh optr
material region=2


# Solution - use gummel newton for initial then switch to full newton

method gummel newton
solve  vgate=0
save outfile=mesfetex02_1.str
tonyplot  mesfetex02_1.str -set mesfetex02_0.set
solve vdrain=0.025 vstep=0.025 vfinal=0.1 name=drain


method  newton
solve vdrain=0.2 vstep=0.1 vfinal=0.6 name=drain
solve vdrain=0.8 vstep=0.2 vfinal=3 name=drain


# Small signal ac analysis with s-parameter calculation
log outf=mesfetex02_ac.log s.param inport=gate outport=drain width=100
solve ac.analysis direct frequency=1.e9 fstep=2.e9 nfsteps=20


tonyplot mesfetex02_ac.log -set mesfetex02_1.set
tonyplot mesfetex02_ac.log -set mesfetex02_2.set


quit
```

PROBLEMS FOR CHAPTER 5

1. In this exercise we will examine the variation of the energy bands and the quasi-Fermi level in a pn-diode with applied bias. The p-side doping of the Si diode is $N_A = 10^{16}$ cm^{-3} and the n-side doping is $N_D = 10^{16}$ cm^{-3}. The length of the p-side and the n-side region is taken to be 1 μm. Plot the following variables:

   a) Conduction band, valence band, and the intrinsic level variation versus position for applied bias VANODE $= -0.6$, 0, and 0.6 V.

   b) Quasi-Fermi level variation for the above voltage bias conditions.

2. Consider a MOS capacitor structure found in conventional MOSFET devices. The thickness of the oxide region equals 4 nm and the substrate is p-type with doping N_A.

a) Assume that $N_A = 10^{17}$ cm^{-3}. Plot the conduction band profile under equilibrium conditions assuming aluminum, n^+-polysilicon and p^+-polysilicon gate.

b) Vary the gate voltage from -2 to 2 V and calculate the high-frequency CV curves using $f = 1$ MHz. How does the change in the type of the gate electrode (aluminum, versus n^+-polysilicon, versus p^+-polysilicon) reflects on the HF CV-curves.

c) Assume aluminum gate and plot the HF CV-curves for $f = 1$ MHz. How does the change in substrate doping reflect itself on the HF CV-curves. Support your reasoning with a physical model. Assume that $N_A = 10^{16}$, 10^{17}, and 10^{18} cm^{-3}.

3. Consider a 25 nm MOSFET device structure with SiO$_2$ oxide thickness of 1.2 nm and n^+-polysilicon gate. The doping of the source and drain n^+-regions equals 10^{20} cm^{-3}. The depth of the n^+ source and drain regions is 20 nm. Via simulations estimate the necessary substrate doping to ensure proper device operation and to avoid the punch-through effect while maximizing device cutoff frequency. How much current drop we have due to series resistance effects if the doping of the source and drain regions is reduced to 10^{19} cm^{-3}?

4. Examine the differences in the operation of conventional MOSFET device and a SOI MOSFET device. Assume identical device parameters. How do the results differ when using the energy balance instead of the simple drift–diffusion model. Explain your results via physical reasoning.

5. Using the cutline feature of TonyPlot repeat the simulations given in Problem 4 and plot the potential profile across the structure. Is there any difference in the potential profile variation from source to drain in a PD and a FD device. If so, explain the differences observed using physical reasoning.

6. a) Start from the general definition of the h-parameters in the common-base configuration and derive the corresponding ones in the common-emitter configuration. In your derivations first relate the parameters h_{11}, h_{12}, h_{21}, and h_{22} from Figure 5.26(a) to the ones shown in Figure 5.26(b) for the case when the base spreading resistance $r_{bb'} = 0$.

b) How will the h-parameters in both configurations change with the addition of the base spreading resistance $r_{bb'}$? You can arrive at approximate expressions as well considering the smallness of some of the h-parameters.

7. You are given an input deck for simulating a BJT. Run this input deck, write appropriate statements where necessary (mainly extract statements), and provide the following results:

a) Material composition and doping profiles of this device structure and the mesh used in the simulations (provide tree separate plots for clarity).

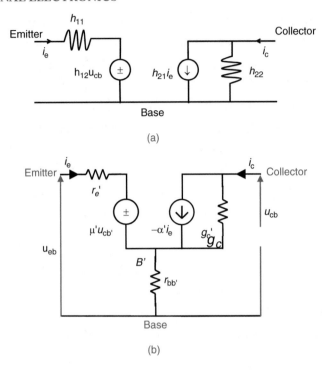

FIGURE 5.26:

b) Variation of the parameter g_m from the Giacolleto equivalent circuit model of the BJT transistor in the common-emitter configuration (shown in Figure 5.27), with the collector current, for fixed values of V_{CE}.

c) Variation of the parameters $g_{b'e}$, $g_{b'c}$, and g_{ce} with the collector current for fixed V_{CE}.

d) Variation of the capacitances $C_{b'e}$ and $C_{b'c}$ with the collector current for fixed V_{CE}.

e) The variation of the cutoff frequency f_T, using the extract model provided in Silvaco and using the approximate expressions derived in class.

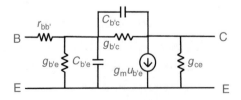

FIGURE 5.27: Giacolleto small-signal equivalent circuit model. Assume $r_{bb'} = 0$ in this case

For cases (b–e), assume that the voltage V_{CE} is varied between 2 and 6 V, in 0.5 V increments. This will give you a family of curves, which can allow you to obtain the variation of each of the above parameters as a function of V_{CE} for fixed I_C. As already given in the input deck, the frequency used for obtaining the small-signal AC parameters is already set to 1 MHz.

8. a) For the description of two-port networks, instead of using the S-parameter set, one can also use the so-called T-parameter set that relates the input to the output variables (T stands for transmission). Given the following general definitions for the S- and T-parameters of a two-port network, express the elements of the T-matrix in terms of the elements of the S-matrix.

$$\begin{bmatrix} b_1 \\ b_2 \end{bmatrix} = \begin{bmatrix} S_{11} & S_{12} \\ S_{21} & S_{22} \end{bmatrix} \begin{bmatrix} a_1 \\ a_2 \end{bmatrix}; \begin{bmatrix} a_1 \\ b_1 \end{bmatrix} = \begin{bmatrix} T_{11} & T_{12} \\ T_{21} & T_{22} \end{bmatrix} \begin{bmatrix} b_2 \\ a_2 \end{bmatrix}$$

b) In most cases it is lot easier to measure the S-parameters under conjugate matching conditions, which do not require short or open-circuits, and calculate the Y-parameters. Therefore, the relationship between the two is an important one. Here you are asked the opposite problem, i.e., given the Y-parameters, you need to find the corresponding expressions for the elements of the S-matrix, i.e., to express the S-parameters in terms of the Y-parameters.

CHAPTER 6

Particle-Based Device Simulation Methods

In the previous chapters we considered continuum methods of describing transport in semiconductors, specifically the drift-diffusion and hydrodynamic models, which are derived from moments of the semiclassical Boltzmann Transport Equation (BTE). As approximations to the BTE, it is expected that at some limit, such approaches become inaccurate, or fail completely. Indeed, one can envision that, as physical dimensions are reduced, at some level a continuum description of current breaks down, and the granular nature of the individual charge particles constituting the charge density in the active device region becomes important. For example, in the ultimate limit of very small geometries at low temperatures, numerous experiments over the past two decades have clearly demonstrated the individual motion of single electrons in so-called single electron transistors.

The microscopic simulation of the motion of individual particles in the presence of the forces acting on them due to external fields as well as the internal fields of the crystal lattice and other charges in the system has long been popular in the chemistry community, where *molecular dynamics* simulation of atoms and molecules have long been used to investigate the thermodynamic properties of liquids and gases. In solids such as semiconductors and metals, transport is known to be dominated by random scattering events due to impurities, lattice vibrations, etc., which randomize the momentum and energy of charge particles in time. Hence, stochastic techniques to model these random scattering events are particularly useful in describing transport in semiconductors, in particular the *Monte Carlo* method.

The Ensemble Monte Carlo techniques have been used for well over 30 years as a numerical method to simulate nonequilibrium transport in semiconductor materials and devices and has been the subject of numerous books and reviews [82–84]. In application to transport problems, a random walk is generated using the random number generating algorithms common to modern computers, to simulate the stochastic motion of particles subject to collision processes. This process of random walk generation is part of a very general technique used to evaluate integral equations and is connected to the general random sampling technique used in the evaluation of multidimensional integrals [85].

The basic technique as applied to transport problems is to simulate the free particle motion (referred to as the free flight) terminated by instantaneous random scattering events. The Monte Carlo algorithm consists of generating random free-flight times for each particle, choosing the type of scattering occurring at the end of the free flight, changing the final energy and momentum of the particle after scattering, and then repeating the procedure for the next free flight. Sampling the particle motion at various times throughout the simulation allows for the statistical estimation of physically interesting quantities such as the single particle distribution function, the average drift velocity in the presence of an applied electric field, the average energy of the particles, *etc.* By simulating an *ensemble* of particles, representative of the physical system of interest, the nonstationary time-dependent evolution of the electron and hole distributions under the influence of a time-dependent driving force may be simulated.

This particle-based picture, in which the particle motion is decomposed into free flights terminated by instantaneous collisions, is basically the same approximate picture underlying the derivation of the semiclassical BTE, discussed in Chapter 2. In fact, it may be shown that the one-particle distribution function obtained from the random walk Monte Carlo technique satisfies the BTE for a homogeneous system in the long-time limit [86]. This semiclassical picture breaks down when quantum mechanical effects become pronounced, and one cannot unambiguously describe the instantaneous position and momentum of a particle, a subject which we will comment on later.

In the following, we develop the standard Monte Carlo algorithm used to simulate charge transport in semiconductors, and how useful quantities are calculated from this. Special considerations of multicarrier effects are then considered, such as carrier–carrier scattering, and impact ionization. We then discuss how this basic model for charge transport within the BTE is self-consistently solved with the appropriate field equations to perform particle-based device simulation.

6.1   FREE-FLIGHT GENERATION

In the Monte Carlo method, particle motion is assumed to consist of free flights terminated by instantaneous scattering events, which change the momentum and energy of the particle after scattering. So the first task is to generate free flights of random time duration for each particle. To simulate this process, the probability density, $P(t)$, is required, in which $P(t)dt$ is the joint probability that a particle will arrive at time t without scattering after a previous collision occurring at time $t = 0$, and then suffer a collision in a time interval dt around time t. The probability of scattering in the time interval dt around t may be written as $\Gamma[\mathbf{k}(t)]dt$, where $\Gamma[\mathbf{k}(t)]$ is the t scattering rate of an electron or hole of wavevector $\mathbf{k}$. The total scattering rate, $\Gamma[\mathbf{k}(t)]$, represents the sum of the contributions from each individual scattering mechanism, which may be calculated quantum mechanically using Fermi's golden rule, Eq. (2.36), as discussed in Chapter 2. The implicit dependence of $\Gamma[\mathbf{k}(t)]$ on time reflects the change in $\mathbf{k}$ due to acceleration by internal and external fields. For electrons subject to time independent electric and magnetic fields the time evolution of $\mathbf{k}$ between collisions is

$$\mathbf{k}(t) = \mathbf{k}(0) - \frac{e(\mathbf{E} + \mathbf{v} \times \mathbf{B})t}{\hbar}, \tag{6.1}$$

where $\mathbf{E}$ is the electric field, $\mathbf{v}$ is the electron velocity, and $\mathbf{B}$ is the magnetic flux density. In terms of the scattering rate, $\Gamma[\mathbf{k}(t)]$, the probability that a particle has not suffered a collision after a time t is given by $\exp(-\int_0^t \Gamma[\mathbf{k}(t')]dt')$. Thus, the probability of scattering in the time interval dt after a free flight of time t may be written as the joint probability

$$P(t)dt = \Gamma[\mathbf{k}(t)]\exp\left[-\int_0^t \Gamma[\mathbf{k}(t')]dt'\right]dt. \tag{6.2}$$

Random flight times may be generated according to the probability density $P(t)$ above using, for example, the pseudo-random number generator implicit on most modern computers, which generate uniformly distributed random numbers in the range $[0, 1]$. Using a direct method (see, for example [82]), random flight times sampled from $P(t)$ may be generated according to

$$r = \int_0^{t_r} P(t)dt, \tag{6.3}$$

where r is a uniformly distributed random number and t_r is the desired free-flight time. Integrating Eq. (6.3) with $P(t)$ given by Eq. (6.2) above yields

$$r = 1 - \exp\left[-\int_0^{t_r} \Gamma[\mathbf{k}(t')]dt'\right]. \tag{6.4}$$

Since $1 - r$ is statistically the same as r, Eq. (6.4) may be simplified to

$$-\ln r = \int_0^{t_r} \Gamma[\mathbf{k}(t')]dt'. \tag{6.5}$$

Equation (6.5) is the fundamental equation used to generate the random free-flight time after each scattering event, resulting in a random walk process related to the underlying particle distribution function. If there is no external driving field leading to a change of $\mathbf{k}$ between scattering events (for example in ultrafast photoexcitation experiments with no applied bias), the time dependence vanishes, and the integral is trivially evaluated. In the general case where this simplification is not possible, it is expedient to introduce the so-called self-scattering method [87], in which we introduce a fictitious scattering mechanism whose scattering rate always adjusts itself in such a way that the total (self-scattering plus real-scattering) rate is a constant in time

$$\Gamma = \Gamma\left[\mathbf{k}\left(t'\right)\right] + \Gamma_{\text{self}}\left[\mathbf{k}\left(t'\right)\right],\tag{6.6}$$

where $\Gamma_{\text{self}}[\mathbf{k}(t')]$ is the self-scattering rate. The self-scattering mechanism itself is defined such that the final state before and after scattering is identical. Hence, it has no effect on the free-flight trajectory of a particle when selected as the terminating scattering mechanism, yet results in the simplification of Eq. (6.5) such that the free flight is given by

$$t_r = -\frac{1}{\Gamma}\ln r.\tag{6.7}$$

The constant total rate (including self-scattering) Γ, must be chosen at the start of the simulation interval (there may be multiple such intervals throughout an entire simulation) so that it is larger than the maximum scattering encountered during the same time interval. In the simplest case, a single value is chosen at the beginning of the entire simulation (constant gamma method), checking to ensure that the real rate never exceeds this value during the simulation. Other schemes may be chosen that are more computationally efficient, and which modify the choice of Γ at fixed time increments [88].

6.2   FINAL STATE AFTER SCATTERING

The algorithm described above determines the random free-flight times during which the particle dynamics is treated semiclassically according to Eq. (6.1). For the scattering process itself, we need the type of scattering (i.e., impurity, acoustic phonon, photon emission, etc.) which terminates the free flight, and the final energy and momentum of the particle(s) after scattering, which were discussed in more detail in Section 2.7. The type of scattering which terminates the free flight is chosen using a uniform random number between 0 and Γ, and using this pointer to select among the relative total scattering rates of all processes including self-scattering at the final energy and momentum of the particle

$$\Gamma = \Gamma_{\text{self}}\left[n, \mathbf{k}\right] + \Gamma_1\left[n, \mathbf{k}\right] + \Gamma_2\left[n, \mathbf{k}\right] + \ldots \Gamma_N\left[n, \mathbf{k}\right],\tag{6.8}$$

with n the band index of the particle (or subband in the case of reduced-dimensionality systems), and $\mathbf{k}$ the wavevector at the end of the free flight. This process is illustrated schematically in Figure 6.1.

Once the type of scattering terminating the free flight is selected, the final energy and momentum (as well as band or subband) of the particle due to this type of scattering must be selected. For elastic scattering processes such as ionized impurity scattering, the energy before

Self	
5	$\Gamma_1 + \Gamma_2 + \Gamma_3 + \Gamma_4 + \Gamma_5$
4	$\Gamma_1 + \Gamma_2 + \Gamma_3 + \Gamma_4$
3	$\Gamma_1 + \Gamma_2 + \Gamma_3$
2	$\Gamma_1 + \Gamma_2$
1	$\Gamma_1(E(t_r))$

Γ $r\Gamma$

FIGURE 6.1: Selection of the type of scattering terminating a free flight in the Monte Carlo algorithm

and after scattering is the same. For the interaction between electrons and the vibrational modes of the lattice described as quasi-particles known as phonons, electrons exchange finite amounts of energy with the lattice in terms of emission and absorption of phonons. For determining the final momentum after scattering, the scattering rate, $\Gamma_j[n, \mathbf{k}; m, \mathbf{k}']$ given in Eq. (2.35) for the jth scattering mechanism is needed, where n and m are the initial and final band indices, and $\mathbf{k}$ and $\mathbf{k}'$ are the particle wavevectors before and after scattering. Defining a spherical coordinate system as shown in Figure 6.2 around the initial wavevector $\mathbf{k}$, the final wavevector $\mathbf{k}'$ is specified by $|\mathbf{k}'|$ (which depends on conservation of energy) as well as the azimuthal and polar angles,

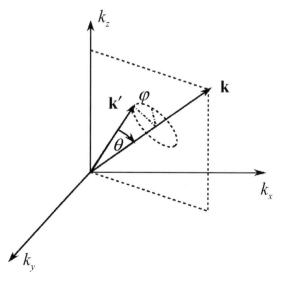

FIGURE 6.2: Coordinate system for determining the final state after scattering

φ and θ around $\mathbf{k}$. Typically, the scattering rate, $\Gamma_j[n, \mathbf{k}; m, \mathbf{k}']$, only depends on the angle θ between $\mathbf{k}$ and $\mathbf{k}'$. Therefore, φ may be chosen using a uniform random number between 0 and 2π (i.e., $2\pi r$), while θ is chosen according to the angular dependence for scattering arising from $\Gamma_j[n, \mathbf{k}; m, \mathbf{k}']$. If the probability for scattering into a certain angle $P(\theta)d\theta$ is integrable, then random angles satisfying this probability density may be generated from a uniform distribution between 0 and 1 through inversion of Eq. (6.3). Otherwise, a rejection technique (see, for example, [82, 83]) may be used to select random angles according to $P(\theta)$.

The rejection technique for sampling a random variable over some interval corresponds to choosing a maximum probability density (referred to here as a maximizing function) that is integrable in terms of Eq. (6.3) (for example a uniform or constant probability), and is always greater than or equal to the actual probability density of interest. A sample value of the random variable is then selected using a uniform number between 0 and 1, and then applying Eq. (6.3) to the maximizing function to select a value of the random variable analytically according to the probability density of the maximizing function. To now sample according to the desired probability density, a second random number is picked randomly between 0 and the value of the maximizing function at the value of the random variable chosen. If the value of this random number is less than the true value of the probability density (i.e., lies below it) at that point, the sampled value of the random variable is "selected." If it lies above, it is "rejected," and the process repeated until one satisfying the condition of selected is generated. In choosing random samples via this technique, one then samples according to the desired probability density.

The idea of the rejection technique is most easily illustrated with an example. Let us consider the polar angular dependence of the scattering rate of electrons in a prototypical polar semiconductor such as GaAs, due to polar optical phonon scattering. The probability density for scattering into an angle θ, relative to the original trajectory of travel just before scattering, is given by

$$P(\theta)\,d\theta = \frac{\Gamma_{\mathrm{POP}}(\theta)\,d\theta}{\int_0^\pi \Gamma_{\mathrm{POP}}(\theta)\,d\theta} \sim \frac{\sin(\theta)\,d\theta}{\left(E + E' - 2\sqrt{EE'}\cos\theta\right)}, \tag{6.9}$$

where $\Gamma_{\mathrm{POP}}(\theta)$ is the scattering rate into a small angle $d\theta$ around the angle θ, E is the energy of the particle before scattering, and $E' = E \pm \hbar\omega_{\mathrm{o}}$ is the final energy corresponding to the emission (lower sign) or absorption (upper sign) of an optical phonon of energy $\hbar\omega_{\mathrm{o}}$. The integral in the denominator is for normalization. The angular dependence of the scattering rate is plotted in Figure 6.3.

In using the rejection method to generate random angles between 0 and π, we choose for simplicity a constant maximizing function, $\Gamma_{\mathrm{POP}}(\theta) = \Gamma_{\mathrm{max}}$. Then, including the normalization, integration of Eq. (6.3) gives $\theta_r = r\pi$, i.e., simply a uniformly distributed random

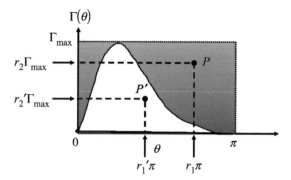

FIGURE 6.3: Use of the rejection method in determining the final angle after scattering for polar optical phonon scattering

angle between 0 and π. Suppose that the first random number we generate is r_1 giving the first random angle $\theta_1 = r_1\pi$. A second random number is generated, r_2, and then multiplied by the value of the maximizing function at $\theta_1 = r_1\pi$. For the particular case visualized in Figure 6.3, P lies above the actual function, $\Gamma_{POP}(\theta_1)$, and therefore is rejected. A second pair of random numbers is generated by the primed coordinates, which leads to the point P', which does lie below $\Gamma_{POP}(\theta_1')$, and is therefore accepted as a valid sample. One can qualitatively see from this example that random angles are preferentially selected under the region of the probability distribution where it is greatest.

6.3   ENSEMBLE MONTE CARLO SIMULATION

The basic Monte Carlo algorithm described in the previous sections may be used to track a single particle over many scattering events in order to simulate the steady-state behavior of a system. However, for improved statistics over shorter simulation times, and for transient simulation, the preferred technique is the use of a *synchronous ensemble* of particles, in which the basic Monte Carlo algorithm is repeated for each particle in a ensemble representing the (usually larger) system of interest until the simulation is completed. Since there is rarely an identical correspondence between the number of simulated charges, and the number of actual particles in a system, each particle is really a *super-particle*, representing a finite number of real particles. The corresponding charge of the particle is weighted by this super-particle number. Figure 6.4 illustrates an ensemble Monte Carlo simulation in which a fixed time step, Δt, is introduced over which the motion of all the carriers in the system is synchronized. The squares illustrate random, instantaneous, scattering events, which may or may not occur during a given time step. Basically, each carrier is simulated only up to the end of the time step, and then the next particle in the ensemble is treated. Over each time step, the motion of each particle in

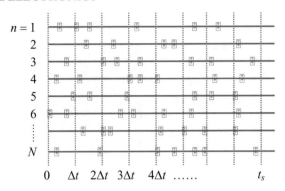

FIGURE 6.4: Ensemble Monte Carlo simulation in which a time step, Δt, is introduced over which the motion of particles is synchronized. The squares represent random scattering events

the ensemble is simulated independent of the other particles. Nonlinear effects such as carrier–carrier interactions or the Pauli exclusion principle are then updated at each times step, as discussed in more detail below.

The nonstationary one-particle distribution function and related quantities such as drift velocity, valley or subband population, etc., are then taken as averages over the ensemble at fixed time steps throughout the simulation. For example, the drift velocity in the presence of the field is given by the ensemble average of the component of the velocity at the nth time step as

$$\bar{v}_z(n\Delta t) \cong \frac{1}{N}\sum_{j=1}^{N} v_z^j(n\Delta t), \qquad (6.10)$$

where N is the number of simulated particles and j labels the particles in the ensemble. This equation represents an estimator of the true velocity, which has a standard error given by

$$s = \frac{\sigma}{\sqrt{N}}, \qquad (6.11)$$

where σ^2 is the variance which may be estimated from [85]

$$\sigma^2 \cong \frac{N}{N-1}\left\{\frac{1}{N}\sum_{j=1}^{N}\left(v_z^j\right)^2 - \bar{v}_z^2\right\}. \qquad (6.12)$$

Similarly, the distribution functions for electrons and holes may be tabulated by counting the number of electrons in cells of k-space. From Eq. (6.11), we see that the error in estimated average quantities decreases as the square root of the number of particles in the ensemble, which necessitates the simulation of many particles. Typical ensemble sizes for good statistics are in the range of $10^4 - 10^5$ particles. Variance reduction techniques to decrease the standard error given by Eq. (6.11) may be applied to enhance statistically rare events such as impact ionization or electron–hole recombination [83].

An overall flowchart of a typical Ensemble Monte Carlo (EMC) simulation is illustrated in Figure 6.4. After initialization of run parameters, there are two main loops, and outer one which advances the time step by increments of ΔT until the maximum time of the simulation is reached, and an inner loop over all the particles in the ensemble (N), where the Monte Carlo algorithm is applied to each particle individually over a given time step.

As an example of the calculated results for the EMC algorithm illustrated in Figures 6.5 and 6.6 show the calculated velocity in the direction of the electric field (drift velocity) versus time for GaAs at 300 K for various electric fields, in which the ensemble of carrier is initially in equilibrium, and then a constant electric field is abruptly turned on at zero. The model used here is a nonparabolic three valley model, consisting of a central valley surrounded by satellite valleys in the X and L directions. Scattering mechanisms included are polar optical phonon scattering, acoustic deformation potential scattering, intervalley nonpolar optical scattering, and ionized impurity scattering (impurity concentration $= 1.0 \times 10^{14}$ cm^{-3}). The first thing

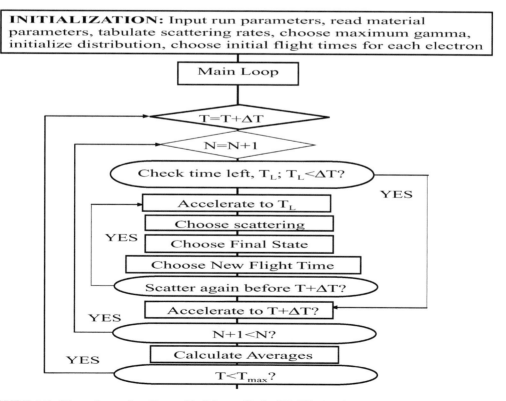

FIGURE 6.5: Flow chart of an Ensemble Monte Carlo (EMC) simulation

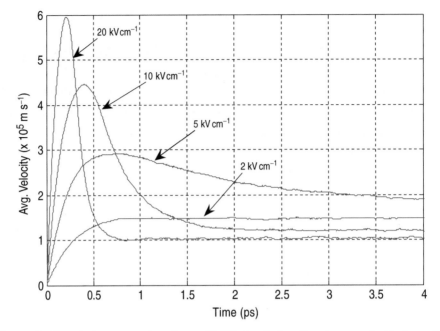

FIGURE 6.6: Drift velocity versus time in an EMC simulation for electrons in GaAs at 300 K for various electric fields

to note is that there is a transient period which may last over several picoseconds, before the carriers reach a steady-state situation. For very short times, the motion of particles is almost ballistic (free of scattering) as they accelerate freely in time. As scattering begins to occur, the carrier acceleration slows, and the velocity reaches a peak (overshoot) before settling to a steady state. The overshoot velocity becomes more pronounced at higher fields, and is related to differences in the momentum and energy relaxation times in the system associated with scattering as discussed in the preceding chapters, as well as intervalley transfer which occurs when carriers are accelerated high enough in energy to overcome the energy difference of the valleys (approximately 0.28 eV in GaAs).

Figure 6.7 shows the steady-state drift velocity versus electric field, calculated by waiting until the electron velocity in Figure 6.6 reaches steady state, and then performing averages both in time and over the ensemble to calculate the stationary velocity for a given field. As can be observed, the velocity versus field is initially linear in the field, with the slope given by the low-field mobility of GaAs. At the peak of the velocity–field curve, the velocity saturates and then decreases, due to the transfer of carriers from the higher mobility central valley, to the lower mobility satellite valleys. This mechanism is responsible for a region of negative resistance and corresponding Gunn oscillations due to the ensuing instability associate with negative resistance.

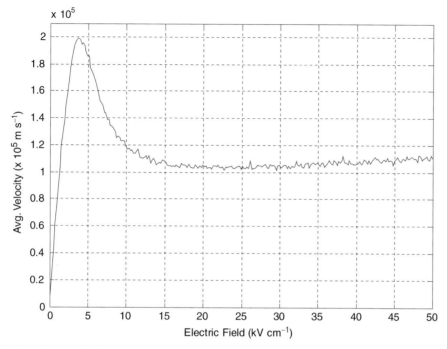

FIGURE 6.7: Average drift velocity versus electric field for Bulk GaAs at 300 K

6.4   MULTICARRIER EFFECTS

Multiparticle effects relate to the interaction between particles in the system, which is a nonlinear effect when viewed in the context of the BTE, due to the dependence of such effects on the single particle distribution function itself. Particle–particle interactions are important in Monte Carlo simulation in establishing or relaxing to an equilibrium distribution function characterized by a Maxwell–Boltzman distribution for nondegenerate situations, or a Fermi–Dirac when proper account for the Pauli exclusion principle is included. Most algorithms developed to deal with such effects essentially linearize the BTE by using the previous value of the distribution function to determine the time evolution of a particle over the successive time step. Multicarrier effects may range from simple consideration of the Pauli exclusion principle (which depends on the exact occupancy of states in the system), to single particle and collective excitations in the system. Inclusion of carrier–carrier interactions in Monte Carlo simulation has been an active area of research for quite some time and is briefly discussed below. Another carrier–carrier effect that is of considerable importance when estimating leakage currents for example in MOSFET devices, is impact ionization, which is a pure generation process involving three particles (two electrons and a hole or two holes and an electron). The latter is also discussed below.

6.4.1   Pauli Exclusion Principle

The Pauli exclusion principle requires that the bare scattering rate be modified by a factor $1 - f_m(\mathbf{k}')$ in the collision integral of the BTE, where $f_m(\mathbf{k}')$ is the one-particle distribution function for the state $\mathbf{k}'$ in band (subband) m after scattering. Since the net scattering rate including the Pauli exclusion principle is always less than the bare scattering rate, a self-scattering rejection technique may be used in the Monte Carlo simulation as proposed by Bosi and Jacoboni [89] for one particle simulation and extended by Lugli and Ferry [90] for EMC. In the self-scattering rejection algorithm, an additional random number r is generated (between 0 and 1), and this number is compared to $f_m(\mathbf{k}')$, the occupancy of the final state (which is also between 0 and 1 when properly normalized for the numerical $\mathbf{k}$-space discretization). If r is greater than $f_m(\mathbf{k}')$, the scattering is accepted and the particle's momentum and energy are changed. If this condition is not satisfied, the scattering is rejected, and the process is treated as a self-scattering event with no change of energy or momentum after scattering. Through this algorithm, no scattering to this state can occur if the state is completely full.

6.4.2   Carrier–Carrier Interactions

Carrier–carrier scattering, may be treated as a scattering process within the Monte Carlo algorithm on the same footing as other mechanisms. In the simplest case of bulk electrons in a single parabolic conduction band, the process may be treated as a binary collision where the scattering rate for a particle of wavevector $\mathbf{k}_0$ due to all the other particles in the ensemble is given by [91]

$$\Gamma_{ee}(\mathbf{k}_0) = \frac{nm_n e^4}{4\pi\hbar^3 \varepsilon^2 \beta^2} \int d\,\mathbf{k} f(\mathbf{k}) \frac{|\mathbf{k} - \mathbf{k}_0|}{\left(|\mathbf{k} - \mathbf{k}_0|^2 + \beta^2\right)}, \tag{6.13}$$

where $f(\mathbf{k})$ is the one-particle distribution function (normalized to unity), ε is the permittivity, n is the electron density, and β is the screening constant. In deriving Eq. (6.13), one assumes that the two particles interact through a statically screened Coulomb interaction, which ignores the energy exchange between particles in the screening which in itself is a dynamic, frequency-dependent effect. Similar forms have been derived for electrons in 2D [92, 93] and 1D [94], where carrier–carrier scattering leads to intersubband as well as intra-subband transitions. Since the scattering rate in Eq. (6.13) depends on the distribution function of all the other particles in the system, this process represents a nonlinear term as mentioned earlier. One method is to tabulate $f(\mathbf{k})$ on a discrete grid, as is done for the Pauli principle, and then numerically integrate Eq. (6.13) at each time step. An alternate method is to use a self-scattering rejection technique [95], where the integrand excluding $f(\mathbf{k})$ is replaced by its maximum value and taken outside the integral over $\mathbf{k}$. The integral over $f(\mathbf{k})$ is just unity, giving an analytic form used to generate the free flight. Then, the self-scattering rejection technique is used when the final state is chosen to correct for the exact scattering rate compared to this artificial maximum rate, similar to the algorithm used for the Pauli principle.

The treatment of intercarrier interactions as binary collisions above neglects scattering by collective excitations such as plasmons or coupled plasmon–phonon modes. These effects may have a strong influence on carrier relaxation, particularly at high carrier density. One approach is to make a separation of the collective and single particle spectrum of the interacting many-body Hamiltonian, and treat them separately, i.e., as binary collisions for the single particle excitations, and as electron–plasmon scattering for the collective modes [96]. Another approach is to calculate the dielectric response within the random phase approximation, and associate the damping given by the imaginary part of the inverse dielectric function with the electron lifetime [97].

A semiclassical approach to carrier–carrier interaction, which is fully compatible with the Monte Carlo algorithm, is the use of Molecular Dynamics [98], in which carrier–carrier inter-action is treated continuously in real space during the free-flight phase through the Coulomb force of all the particles. A very small time step is required when using Molecular Dynamics to account for the dynamic distribution of the system. A time step on the order of 0.5 fs is often sufficiently small for this purpose. The small time step assures that the forces acting on the particles during the time of flight are essentially constant, that is $f(t) \cong f(t + \Delta t)$, where $f(t)$ is the single particle distribution function.

Using Newtonian kinematics, we can write the real space trajectories of each particle as

$$\mathbf{r}(t + \Delta t) = \mathbf{r}(t) + \mathbf{v}\Delta t + \frac{1}{2}\frac{\mathbf{F}(t)}{m}\Delta t^2 \qquad (6.14)$$

and

$$\mathbf{v}(t + \Delta t) = \mathbf{v}(t) + \frac{\mathbf{F}(t)}{m}\Delta t. \qquad (6.15)$$

Here, $\mathbf{F}(t)$ is the force arising from the applied field as well as that of the Coulomb interactions. We can write $\mathbf{F}(t)$ as

$$\mathbf{F}(t) = q\left[\mathbf{E} - \sum_i \nabla\varphi(\mathbf{r}_i(t))\right], \qquad (6.16)$$

where $q\mathbf{E}$ is the force due to the applied field and the summation is the interactive force due to all particles separated by distance $\mathbf{r}_i$, with $\varphi(\mathbf{r}_i)$ the electrostatic potential. As in Monte Carlo simulation, one has to simulate a finite number of particles due to practical computational limitations on execution time. In real space, this finite number of particles corresponds to a particular simulation volume given a certain density of carriers, $V = N/n$, where n is the density.

Since the carriers can move in and out of this volume, and since the Coulomb interaction is a long-range force, one must account for the region outside V by periodically replicating the simulated system. The contributions due to the periodic replication of the particles inside V in cells outside has a closed form solution in the form of an Ewald sum [99], which gives a linear as well as $1/r^2$ contribution to the force. The equation for the total force acting on a given particle due to all other particles in the Molecular Dynamics technique then becomes

$$\mathbf{F} = \frac{-e^2}{4\pi\varepsilon} \sum_i^N \left(\frac{1}{\mathbf{r}_i^2}\mathbf{a}_i + \frac{2\pi}{3V}\mathbf{r}_i \right). \tag{6.17}$$

The above equation is easily incorporated in the standard Monte Carlo simulation discussed up to this point. At every time step the forces on each particle due to all the other particles in the system are calculated from Eq. (6.17). From the forces, an interactive electric field is obtained which is added to the external electric field of the system to couple the Molecular Dynamics to the Monte Carlo.

The inclusion of the carrier–carrier interactions in the context of particle-based device simulations is discussed later in Section 6.5.2. The main difficulty in treating this interaction term in device simulations arises from the fact that the long-range portion of the carrier–carrier interaction is included via the numerical solution of the quasi-static Poisson equation. Under these circumstances, special care has to be taken when incorporating the short-range portion of this interaction term to prevent double counting of the force.

6.4.3 Band to Band Impact Ionization

Another carrier–carrier scattering process is that of impact ionization, in which an energetic electron (or hole) has sufficient kinetic energy to create an electron–hole pair. Impact ionization therefore leads to the process of carrier multiplication. This process is critical for example in the avalanche breakdown of semiconductor junctions, and is a detrimental effect in short channel MOS devices in terms of excess substrate current and decreased reliability.

The ionization rate of valence electrons by energetic conduction band electrons is usually described by Fermi's rule in which a screened Coulomb interaction is assumed between the two particles, as discussed earlier in this section, where screening is described by an appropriate dielectric function such as that proposed by Levine and Louie [100]. In general, the impact ionization rate should be a function of the wavevector of the incident electron, hence of the direction of an electric field in the crystal, although there is still some debate as to the experimental and theoretical evidence. More simply, the energy dependent rate (averaged over all wavevectors on a constant energy shell) may be expressed analytically in the power law form

$$\Gamma_{ii}(E) = P[E - E_{th}]^a, \tag{6.18}$$

where E_{th} is the threshold energy for the process to occur, which is determined by momentum and energy conservation considerations, but minimally is the bandgap of the material itself. P and a are parameters which may be fit to more sophisticated models. The Keldysh formula [101] is derived by expanding the matrix element for scattering close to threshold, which gives $a = 2$, and the constant $P = C/E_{th}^2$, with $C = 1.19 \times 10^{14}$ s^{-1} and assuming a parabolic band approximation,

$$E_{th} = \frac{3 - 2m_v/m_c}{1 - m_v/m_c} E_g, \qquad (6.19)$$

where m_v and m_c are the effective masses of the valence and conduction band respectively, and E_g is the bandgap. More complete full-band structure calculations of the impact ionization rate have been reported for Si [102,103], GaAs [103,104] and wide bandgap materials [105], which are fairly well fit using power law model given in Eq. (6.18).

Within the ensemble Monte Carlo method, the maximum scattering rate is used to generate the free-flight time. The state after scattering of the initial electron plus the additional electron and hole must satisfy both energy and momentum conservation within the Fermi rule model, which is somewhat complicated unless simple parabolic band approximations are made.

6.4.4   Full-Band Particle-Based Simulation

The Monte Carlo algorithm discussed in this section initially evolved during the 1970s and early 1980s using simplified representations of the electronic band structure in terms of a multivalley parabolic or nonparabolic approximation close to band minima and maxima. This simplifies the particle tracking in terms of the E–k relationship and particle motion in real space, and greatly simplifies the calculated scattering rates such that analytical forms may be used. It soon became apparent that for devices where high-field effects are important, or for the correct simulation of high energy processes like impact ionization, the full-band structure of the material is required. Particle-based simulation which incorporates part or all of the band structure directly into the particle dynamics and scattering is commonly referred to as *full-band* Monte Carlo simulation [84].

Typically, the Empirical Pseudopotential Method (EPM) [106] has been utilized in full-band Monte Carlo codes due to the relative simplicity of the calculation, and the plane wave basis which facilitates calculation of some scattering processes. Early full-band codes developed at the University of Illinois utilized the full-band structure for the particle dynamics, but assumed isotropic energy dependent scattering rates using the full-band density of states [84]. This is due to the computational difficulty and memory requirements to store the full k-dependent scattering rates throughout the whole Brillouin zone. Later simulators relaxed this restriction, although often assuming quasi-isotropic rates. Probably the most completely developed full-band code for full-band Monte Carlo device simulation is the DAMOCLES code developed at IBM by Fischetti and Laux [107], which has been used extensively for simulation of a variety of device technologies [108].

The full-band codes above are based on essentially the same algorithm as discussed above, in which a particle scatters based on the total scattering rate, then the type of scattering and the final state after scattering are selected using the full **k**-dependent rates for each mechanism. An alternative approach, referred to as Cellular Monte Carlo [109], stores the entire transition table for the total scattering rate for all mechanisms from every initial state k to every final state k'. Particle scattering is accomplished in a single step, at the expense of large memory consumption (on the order of 2 GB of RAM) necessary to store the necessary scattering tables.

Figure 6.8 shows the calculated steady-state drift velocity and average energy for Si as a function of electric field for the CMC method and the earlier results from DAMOCLES which are essentially the same. In such simulations, steady state is typically reached after 2 ps

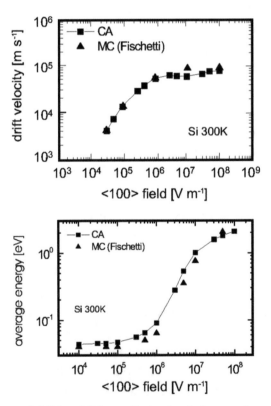

FIGURE 6.8: Comparison of full-band Monte Carlo simulation results using DAMOCLES [107] (triangles) to those using the CMC approach [109]. The upper plot is the steady-state drift velocity and the lower plot the average energy versus electric field

of simulation time, and then averages are calculated over the ensemble and in time for several picoseconds thereafter.

6.5   DEVICE SIMULATION USING PARTICLES

In previous sections of this chapter, we introduced the numerical solution of the BTE using Monte Carlo methods, and showed some results for the simulation of electrons in semiconductors under the influence of a constant electric field. Within an inhomogeneous device structure, however, both the transport dynamics and an appropriate field solver are coupled to each other. For quasi-static situations, as discussed in earlier chapters, the spatially varying fields associated with the potential arising from the numerical solution of Poisson's equation are the driving force accelerating particles in the Monte Carlo phase. Likewise, the distribution of mobile (both electrons and holes) and fixed charges (e.g., donors and acceptors) provides the source of the electric field in Poisson's equation corresponding to the right-hand side of Eq. (3.10). By decoupling the transport portion from the field portion over a small time interval (discussed in more detail below), a convergent scheme is realized in which the Monte Carlo transport phase is self-consistently coupled to Poisson's equation, similar to Gummel's algorithm discussed in Chapter 3. In the following section, a description of Monte Carlo particle-based device simulators is given, with emphasis on the particle–mesh coupling and the inclusion of the short-range Coulomb interactions.

6.5.1   Monte Carlo Device Simulation

As mentioned above, for device simulation based on particles, Poisson's equation is decoupled from the particle motion (described, e.g., by the EMC algorithm) over a suitably small time step, typically less than the inverse plasma frequency corresponding to the highest carrier density in the device. Over this time interval, carriers accelerate according to the frozen field profile from the previous time-step solution of Poisson's equation, and then Poisson's equation is solved at the end of the time interval with the frozen configuration of charges arising from the Monte Carlo phase (see discussion in Ref. [98]). It is important to note that Poisson's equation is solved on a discrete mesh, whereas the solution of charge motion using EMC occurs over a continuous range of coordinate space in terms of the particle position. An illustration of a typical device geometry and the particle mesh scheme is shown in Figure 6.9. Therefore, a particle–mesh (PM) coupling is needed for both the charge assignment and the force interpolation. The size of the mesh and the characteristic time scales of transport set constraints on both the time step and mesh size. We must consider how particles are treated in terms of the boundaries, and how they are injected. Finally, the determination of the charge motion and corresponding terminal currents from averages over the simulation results are necessary in order to calculate the *IV*

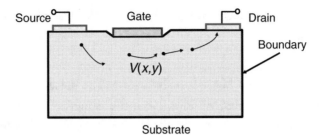

FIGURE 6.9: Schematic diagram of a prototypical three-terminal device where charge flow is described by particles, while the fields are solved on a finite mesh

characteristics of a device. These issues are discussed in detail below, along with some typical simulation results.

6.5.1.1 Time-Step and Grid Size Considerations

As in the case of any time domain simulation, for stable Monte Carlo device simulation, one has to choose the appropriate time step, Δt, and the spatial mesh size (Δx, Δy, and/or Δz). The time step and the mesh size may correlate to each other in connection with the numerical stability. For example, the time step Δt must be related to the plasma frequency

$$\omega_p = \sqrt{\frac{e^2 n}{\varepsilon_s m^*}}, \tag{6.20}$$

where n is the carrier density. From the viewpoint of numerical stability, Δt must be much smaller than the inverse plasma frequency. Since the inverse plasma frequency goes as $1/\sqrt{n}$, the highest carrier density occurring in the modeled device structure corresponds to the smallest time used to estimate Δt. If the material is a multivalley semiconductor, the smallest effective mass encountered by the carriers must be used in Eq. (6.22) as well. For example, in the case of electrons in the central valley of GaAs ($m^* = 0.067 m_o$), a doping of 5×10^{17} cm^{-3} corresponds to $\omega_p \cong 5 \times 10^{13}$; hence, Δt must be smaller than 0.02 ps.

The mesh size for the spatial resolution of the potential is dictated by the spatial charge variation. Hence, one has to choose the mesh size to be smaller than the smallest wavelength of the charge variation. The smallest wavelength is approximately equal to the Debye length (for degenerate semiconductors the relevant length is the Thomas–Fermi wavelength), given by

$$\lambda_D = \sqrt{\frac{\varepsilon_s k_B T}{e^2 n}}. \tag{6.21}$$

Again, due to the inverse dependence of this length on the square root of the density, the highest carrier density modeled should be used to estimate λ_D for stability. Hence, the mesh size must be chosen to be smaller than the Debye length given by Eq. (6.21). Again, for the case of GaAs, with a doping density of 5×10^{17} cm^{-3}, $\lambda_D \cong 6$ nm.

Based on the discussion above, the time step (Δt), and the mesh size (Δx, Δy, an/or Δz) are chosen independently based on the physical arguments given above. However, there are numerical constraints coupling both as well. More specifically, the relation of Δt to the grid size must also be checked by calculating the distance $l_{\max}$, defined as

$$l_{\max} = \mathbf{v}_{\max} \times \Delta t, \qquad (6.22)$$

where $\mathbf{v}_{\max}$ is the maximum carrier velocity, that can be approximated by the maximum group velocity of the electrons in the semiconductor (on the order of 10^8 cm s^{-1}). The distance $l_{\max}$ is the maximum distance the carriers can propagate during Δt. The time step is therefore chosen to be small enough so that $l_{\max}$ is smaller than the spatial mesh size chosen using Eq. (6.21). This constraint arises because for too large of a time step, Δt, there may be substantial change in the charge distribution, while the field distribution in the simulation is only updated every Δt, leading to unacceptable errors in the carrier force.

To illustrate these various constraints, Figure 6.10 illustrates the range of stability for the time step and minimum grid size (adopted from Hockney [98]). The unshaded region corresponds to stable selections of both quantities. The right region is unstable due to the time step being larger than the inverse plasma frequency, whereas the upper region is unstable due to the grid spacing being larger than the Debye length. The velocity constraint bounds the lower side with its linear dependence on time step.

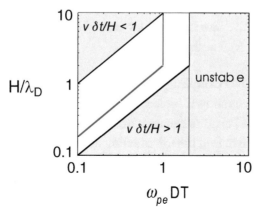

FIGURE 6.10: Illustration of the region of stability (unshaded regions) of the time step, δt, and the minimum grid size, H. ω_{pe} is the plasma frequency corresponding to the maximum carrier density

6.5.1.2 Boundary Conditions for Particles

An issue of importance in particle-based simulation is the real space boundary conditions for the particle part of the simulation. Reflecting or periodic boundary conditions are usually imposed at the artificial boundaries. For Ohmic contacts, they require more careful consideration because electrons (or holes) crossing the source and drain contact regions contribute to the corresponding terminal currents. In order to conserve charge in the device, the electrons exiting the contact regions must be re-injected. Commonly employed models for the contacts include [110]:

- Electrons are injected at the opposite contact with the same energy and wavevector **k**. If the source and drain contacts are in the same plane, as in the case of MOSFET simulations, the sign of **k**, normal to the contact will change. This is an unphysical model, however [111].

- Electrons are injected at the opposite contact with a wavevector randomly selected based upon a thermal distribution. This is also an unphysical model.

- Contact regions are considered to be in thermal equilibrium. The total number of electrons in a small region near the contact are kept constant, with the number of electrons equal to the number of dopant ions in the region. Particles are injected with a velocity weighted by the thermal distribution function. This approximation is most commonly employed in actual particle-based device simulation.

- Another method uses "reservoirs" of electrons adjacent to the contacts. Electrons naturally diffuse into the contacts from the reservoirs, which are not treated as part of the device during the solution of Poisson's equation. This approach gives results similar to the velocity-weighted Maxwellian, but at the expense of increased computational time due to the extra electrons simulated. It is an excellent model employed in some of the most sophisticated particle-based simulators. There are also several possibilities for the choice of the distribution function—Maxwellian, displaced Maxwellian, and velocity-weighted Maxwellian [112].

6.5.1.3 Particle–Mesh (PM) Coupling

The particle–mesh (PM) coupling is broken into four steps: (1) assignment of particle charge to the mesh; (2) solution of Poisson's equation on the mesh; (3) calculation of the mesh-defined forces; and (4) interpolation to find the forces acting on the particle. The charge assignment and force interpolation schemes usually employed in self-consistent Monte Carlo device simulations are the nearest-grid-point (NGP) and the cloud-in-cell (CIC) schemes [113]. Figure 6.11 illustrates both methods. In the NGP scheme, the particle position is mapped into the charge density at the closest grid point to a given particle. This has the advantage of simplicity, but leads to a noisy charge distribution, which may exacerbate numerical instability. Alternately,

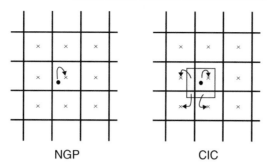

NGP   CIC

FIGURE 6.11: Illustration of the charge assignment based on the nearest-grid-point (NGP) method and the cloud-in-cell (CIC) method

within the CIC scheme a finite volume is associated with each particle spanning several cells in the mesh, and a fractional portion of the charge per particle is assigned to grid points according to the relative volume of the "cloud" occupying the cell corresponding to the grid point. This method has the advantage of smoothing the charge distribution due to the discrete charges of the particle-based method, but may result in an artificial "self-force" acting on the particle, particularly if an inhomogeneous mesh is used.

To better understand the NGP and the CIC scheme, consider a tensor-product mesh with mesh lines x_i, $i = 1, \ldots, N_x$ and y_j, $j = 1, \ldots, N_y$. If the mesh is uniformly spaced in each axis direction, then $(x_{l+1} - x_l) = (x_{l+2} - x_{l+1})$. The permittivities are considered constant within each mesh element and are denoted by ε_{kl}, $k = 1, \ldots, N_x - 1$ and $l = 1, \ldots, N_y - 1$. Define centered finite-differences of the potential ψ in the x- and y-axis at the midpoints of element edges as follows:

$$
\begin{cases}
\Delta^x_{k+\frac{1}{2},l} = -\dfrac{\psi_{k+1,l} - \psi_{k,l}}{x_{k+1} - x_k}, \\
\Delta^y_{k,l+\frac{1}{2}} = -\dfrac{\psi_{k,l+1} - \psi_{k,l}}{y_{l+1} - y_l},
\end{cases}
\tag{6.23}
$$

where the minus sign is included for convenience because the electric field is negative of the gradient of the potential. Consider now a point charge in 2D located at (x, y) within an element $\langle i, j \rangle$. If the restrictions for the permittivity (P) and the tensor-product meshes with uniform spacing in each direction (M) apply, the standard NGP/CIC schemes in two dimensions can be summarized by the following four steps:

1. *Charge assignment to the mesh:* The portion of the charge ρ_L assigned to the element nodes (k, l) is $w_{kl}\rho_L$, $k = i, i + 1$ and $l = j, j + 1$, where w_{kl} are the four charge weights which sum to unity by charge conservation. For the NGP scheme, the node closest to (x, y) receives a weight $w_{kl} = 1$, with the remaining three weights set to zero.

For the CIC scheme, the weights are $w_{ij} = w_x w_y$, $w_{i+1,j} = (1 - w_x) w_y$, $w_{i,j+1} = w_x(1 - w_y)$, and $w_{i+1,j+1} = (1 - w_x)(1 - w_y)$, $w_x = (x_{i+1} - x)/(x_{i+1} - x_i)$ and $w_y = (y_{j+1} - y)/(y_{j+1} - y_j)$.

2. *Solve the Poisson equation:* The Poisson equation is solved by some of the numerical techniques discussed in Appendix A.

3. *Compute forces on the mesh:* The electric field at mesh nodes (k, l) is computed as:
$E^x_{kl} = \left(\Delta^x_{k-\frac{1}{2},l} + \Delta^x_{k+\frac{1}{2},l} \right)/2$ and $E^y_{kl} = \left(\Delta^y_{k,l-\frac{1}{2}} + \Delta^y_{k,l+\frac{1}{2}} \right)/2$, for $k = i, i + 1$ and $l = j, j + 1$.

4. *Interpolate to find forces on the charge:* Interpolate the field to position (x, y) according to $E^x = \sum_{kl} w_{kl} E^x_{kl}$ and $E^y = \sum_{kl} w_{kl} E^y_{kl}$, where $k = i, i + 1, l = j, j + 1$ and the w_{ij} are the NGP or CIC weights from step 1.

The requirements (P) and (M) severely limit the scope of devices that may be considered in device simulations using the NGP and the CIC schemes. Laux [114] proposed a new particle–mesh coupling scheme, namely, the nearest-element-center (NEC) scheme, which relaxes the restrictions (P) and (M). The NEC charge assignment/force interpolation scheme attempts to reduce the self-forces and increase the spatial accuracy in the presence of nonuniformly spaced tensor-product meshes and/or spatially-dependent permittivity. In addition, the NEC scheme can be utilized in one-axis direction (where local mesh spacing is nonuniform) and the CIC scheme can be utilized in the other (where local mesh spacing is uniform). Such hybrid schemes offer smoother assignment/interpolation on the mesh compared to the pure NEC. The new steps of the pure NEC PM scheme are

1' *Charge assignment to the mesh:* Divide the line charge ρ_L equally to the four mesh points of the element $\langle i, j \rangle$.

3' *Compute forces on the mesh:* Calculate the fields $\Delta^x_{i+\frac{1}{2},l}, l = j, j + 1$, and $\Delta^y_{k,j+\frac{1}{2}}, k = i, i + 1$.

4' *Interpolate to find force on the charge:* Interpolate the field according to the following
$E^x = \left(\Delta^x_{i+\frac{1}{2},j} + \Delta^x_{i+\frac{1}{2},j+1} \right)/2$ and $E^y = \left(\Delta^x_{i,j+\frac{1}{2}} + \Delta^x_{i+1,j+\frac{1}{2}} \right)/2$.

The NEC designation derives from the appearance, in step $(1')$ of moving the charge to the center of its element and applying a CIC-like assignment scheme. The NEC scheme involves only one mesh element and its four nodal values of potential. This locality makes the method well-suited to nonuniform mesh spacing and spatially-varying permittivity. The interpolation and error properties of the NEC scheme are similar to the NGP scheme.

6.5.1.4 Real Space Trajectories of Particles

The motion in real space of particles under the influence of electric fields is somewhat more complicated due to the band structure. Recalling the discussion in Chapter 1 the velocity of a particle in real space is related to the E–$\mathbf{k}$ dispersion relation defining the bandstructure as

$$\mathbf{v}(t) = \frac{d\mathbf{r}}{dt} = \frac{1}{\hbar} \nabla_{\mathbf{k}} E(\mathbf{k}(t)), \tag{6.24}$$

$$\frac{d\mathbf{k}}{dt} = \frac{q\mathbf{E}(\mathbf{r})}{\hbar},$$

where the rate of change of the crystal momentum is related to the local electric field acting on the particle through the acceleration theorem expressed by the second equation. In turn, the change in crystal momentum, $\mathbf{k}(t)$, is related to the velocity through the gradient of E with respect to $\mathbf{k}$. If one has to use the full-band structure of the semiconductor, then integration of these equations to find $\mathbf{r}(t)$ is only possible numerically, using for example a Runge–Kutta algorithm [80]. If a three valley model with parabolic bands is used, then the expression is integrable

$$\mathbf{v} = \frac{d\mathbf{r}}{dt} = \frac{\hbar\mathbf{k}}{m^*}; \quad \frac{d\mathbf{k}}{dt} = \frac{q\mathbf{E}(\mathbf{r})}{\hbar}. \tag{6.25}$$

Therefore, for a constant electric field in the x direction, the change in distance along the x direction is found by integrating twice

$$x(t) = x(0) + v_x(0)\, t + \frac{q E_x^0 t^2}{2m^*}. \tag{6.26}$$

6.5.1.5 Simulated Device Behavior

To simulate the steady-state behavior of a device, the system must be initialized in some initial condition, with the desired potentials applied to the contacts, and then the simulation proceeds in a time stepping manner until steady state is reached. This process may take several picoseconds of simulation time, and consequently several thousand time steps based on the usual time increments required for stability. Clearly, the closer the initial state of the system is to the steady-state solution, the quicker the convergence. If one is, for example, simulating the first bias point for a transistor simulation, and has no *a priori* knowledge of the solution, a common starting point for the initial guess is to start out with charge neutrality, i.e., to assign particles randomly according to the doping profile in the device and based on the super-particle charge assignment of the particles, so that initially the system is charge neutral on the average. For two-dimensional device simulation, one should keep in mind that each particle actually represents a rod of charge into the third dimension. Subsequent simulations at the same device at different bias conditions can use the steady-state solution at the previous bias point as a good initial guess. After assigning charges randomly in the device structure, charge is then assigned to each mesh point using the NGP or CIC particle–mesh methods, and Poisson's equation solved. The forces are then interpolated on the grid, and particles are accelerated over the next time step. A flow chart of a typical Monte Carlo device simulation is shown in Figure 6.12.

As the simulation evolves, charge will flow in and out of the contacts, and depletion regions internal to the device will form until steady state is reached. The charge passing through the contacts at each time step can be tabulated, and a plot of the cumulative charge as a function of time plotted as shown in Figure 6.13. Here an n-channel GaAs metal semiconductor field effect transistor (MESFET) device with a Schottky gate is simulated [115]. The Schottky gate is simply modeled as absorbing all carriers from the semiconductor side that have sufficient energy to overcome the potential barrier there, but they are not re-emitted back into the semiconductor, as opposed to the Ohmic contact case. For the particle dynamics, a simple three valley model for GaAs is used including polar optical phonons, intervalley phonon scattering, and impurity scattering. As can be seen, carriers flow out of the gate and drain, and flow in from the source, until a depletion region is established under Schottky gate after about 4 ps. In steady state, the charge versus time is linear, the slope of the source or drain contacts corresponding to the source drain current, while the gate current is approximately zero.

Figure 6.14 shows the particle distribution in 3D of a MESFET, where the dots indicate the individual simulated particles for two different gate biases. Here the heavily doped MESFET region (shown by the inner box) is surrounded by semi-insulating GaAs forming the rest of the simulation domain. The upper curve corresponds to no net gate bias (i.e., the gate is positively biased to overcome the built-in potential of the Schottky contact), while the lower curve corresponds to a net negative bias applied to the gate, such that the channel is close to pinchoff. One can see the evident depletion of carriers under the gate under the latter conditions.

After sufficient time has elapsed, so that the system is driven into a steady-state regime, one can calculate the steady-state current through a specified terminal. The device current can be determined via two different, but consistent methods. First, by keeping track of the charges entering and exiting each terminal, as was done above, the net number of charges over a period of the simulation can be used to calculate the terminal current. This method, however, is relatively noisy due to the discrete nature of the carriers, and the fact that one is only counting the currents crossing a 2D boundary in the device, which limits the statistics. A second method uses the sum of the carrier velocities in a portion of the device are used to calculate the current. For this purpose, the device is divided into several sections along, for example, the x-axis (from source to drain for the case of a MOSFET or MESFET simulation). The number of carriers and their corresponding velocity is added for each section after each free-flight time step. The total x-velocity in each section is then averaged over several time steps to determine the current for

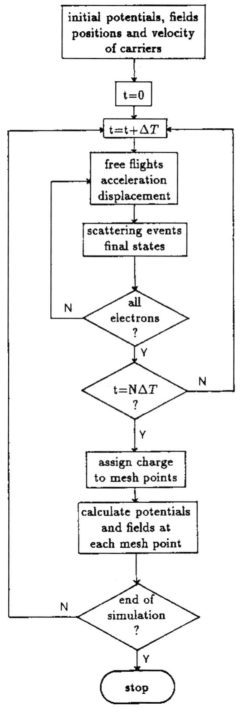

FIGURE 6.12: Flow chart of a typical particle-based device simulation

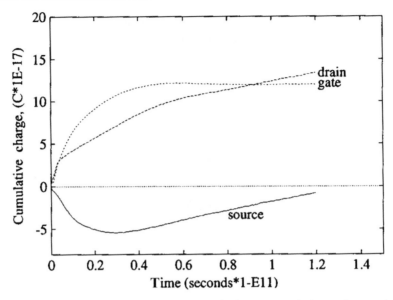

FIGURE 6.13: Plot of the cumulative charge through the source, drain, and gate simulations as a function of time during the simulation for a GaAs MESFET device

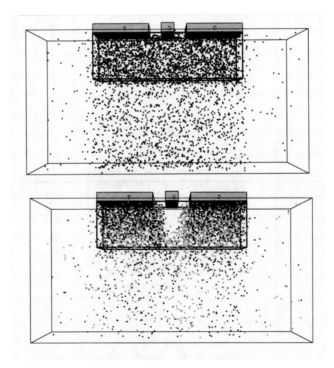

FIGURE 6.14: Example of the particle distribution in a MESFET structure simulated in 3D using an EMC approach. The upper plot is the device with zero gate voltage applied, while the lower is with a negative gate voltage applied, close to pinch-off

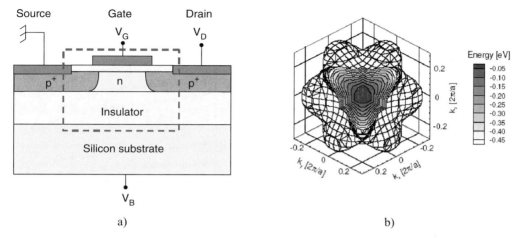

a)   b)

FIGURE 6.15: (a) Cross-section of a p-channel Semiconductor on Insulator (SOI) field effect transistor; (b) constant energy surfaces for the top of the valence band in Ge [116]

that section. The total device current can be determined from the average of several sections, which gives a much smoother result compared to counting the terminal charges. By breaking the device into sections, individual section currents can be compared to verify that the currents are uniform. In addition, sections near the source and drain regions of a MOSFET or a MES-FET may have a high y-component in their velocity and should be excluded from the current calculations.

As a more specific example of calculated current–voltage ($I-V$) characteristics of a device, Figure 6.15 shows the cross-section of a semiconductor on insulator MOS technology, where a buried oxide (BOX) is created by ion-implantation of oxygen or some other process. The top layer is typically Si, although more recently Ge top layers have been of interest. A two-dimenional device simulation based on the computational domain represented by the dashed curve has been performed [116]. Due to the anisotropic and multiband nature of the valence band shown in Figure 6.15b, a full-band Cellular Monte Carlo simulation was performed for hole transport in the channel of these devices.

Figure 6.16 shows the calculated $I_{SD}-V_{SD}$ characteristics for various gate biases relative to the threshold voltage. Here, a 297×65 and 337×65 tensor-product grid was used for the solution of Poisson's equation for the 90 and 130 nm devices, respectively. Eighty thousand super-particles were used in these simulations with a 0.2 fs timestep. The current was calculated through averages over slices in the channel, as discussed earlier. As can be seen, the statistical fluctuation of the calculated curves is relatively small, and systematic differences between Ge on insulator versus Si on insulator are found, due to the higher hole mobility of Ge versus Si.

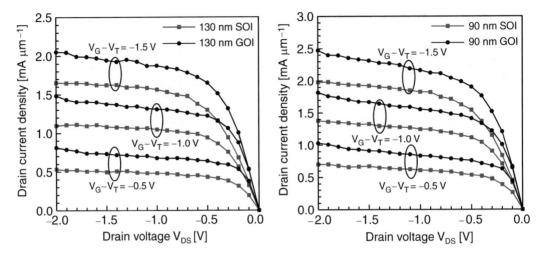

FIGURE 6.16: Calculated *IV* characteristics for the *p*-channel SOI MOSFET structure shown in Figure 6.15 for both Ge on insulator and Si on insulator devices, for 130 and 90 nm gate lengths

6.5.2   Direct Treatment of Interparticle Interaction

In modern deep-submicrometer devices, for achieving optimum device performance and eliminating the so-called punch-through effect, the doping densities must be quite high. This necessitates a careful treatment of the electron–electron (e–e) and electron–impurity (e–i) interactions, an issue that has been a major problem for quite some time. Many of the approaches used in the past have included the short-range portions of the e–e and e–i interactions in the **k**-space portion of the Monte Carlo transport kernel, thus neglecting many of the important inelastic properties of these two interaction terms [117, 118]. An additional problem with this screened scattering approach in devices is that, unlike the other scattering processes, e–e and e–i scattering rates need to be re-evaluated frequently during the simulation process to take into account the changes in the distribution function in time and spatially. The calculation and tabulation of a spatially inhomogeneous distribution function may be highly CPU and memory intensive. Furthermore, ionized impurity scattering is usually treated as a simple two-body event, thus ignoring the multi-ion contributions to the overall scattering potential. A simple screening model is usually used that ignores the dynamical perturbations to the Coulomb fields caused by the movement of the free carriers. To overcome the above difficulties, several authors have advocated coupling of the semiclassical molecular dynamics approach discussed in Section 6.4.2 to the ensemble Monte Carlo approach [119–121]. Simulation of the low-field mobility using such a coupled approach results in excellent agreement with the experimental data for high substrate doping levels [121]. However, it is proven to be quite difficult to incorporate this coupled ensemble Monte Carlo molecular dynamics approach when inhomogeneous charge densities, characteristic of semiconductor devices, are encountered [118,122]. An additional problem with this approach in a typical particle-based device simulation arises from the fact that both the e–e and e–i interactions are already included, at least within the Hartree approximation (long-range carrier–carrier interaction), through the self-consistent solution of the three-dimensional (3D) Poisson equation via the PM coupling discussed in the previous section. The magnitude of the resulting mesh force that arises from the force interpolation scheme, depends upon the volume of the cell, and, for commonly employed mesh sizes in device simulations, usually leads to double-counting of the force.

To overcome the above-described difficulties of incorporation of the short-range e–e and e–i force into the problem, one can follow two different paths. One way is to use the P^3M scheme introduced by Hockney and Eastwood [113]. An alternative to this scheme is to use the corrected-Coulomb approach due to Gross *et al.* [123–126].

6.5.2.1 The P^3M Method

The particle–particle–particle–mesh (P^3M) algorithms are a class of hybrid algorithms developed by Hockney and Eastwood [113]. These algorithms enable correlated systems with long-range forces to be simulated for a large ensemble of particles. The essence of the method is to express the interparticle forces as a sum of two component parts; the short-range part $\mathbf{F}_{sr}$, which is nonzero only for particle separations less than some cutoff radius $\mathbf{r}_e$, and the smoothly varying part $\mathbf{F}$, which has a transform that is approximately band-limited. The total short-range force on a particle $\mathbf{F}_{sr}$ is computed by direct particle–particle (PP) pair force summation, and the smoothly varying part is approximated by the particle–mesh (PM) force calculation.

Two meshes are employed in the P^3M algorithms: the charge-potential mesh and a coarser mesh, the so-called chaining mesh. The charge potential mesh is used at different stages of the PM calculation to store, in turn, charge density values, charge harmonics, potential harmonics and potential values. The chaining mesh is a regular array of cells whose sides have lengths greater than or equal to the cutoff radius $\mathbf{r}_e$ of the short-range force. Associated with each cell of this mesh is an entry in the head-of-chain array. This addressing array is used in conjunction with an extra particle coordinate, the linked-list coordinate, to locate pairs of neighboring particles in the short-range calculation.

The particle orbits are integrated forward in time using the leapfrog scheme

$$\mathbf{x}_i^{n+1} = \mathbf{x}_i^n + \frac{\mathbf{p}_i^{n+1/2}}{m}\Delta t, \tag{6.27}$$

$$\mathbf{p}_i^{n+1/2} = \mathbf{p}_i^{n-1/2} + \left(\mathbf{F}_i + \mathbf{F}_i^{sr}\right)\Delta t. \tag{6.28}$$

The positions $\{\mathbf{x}_i\}$ are defined at integral time-levels and momenta $\{\mathbf{p}_i\}$ are defined at half-integral time-levels. Momenta $\{\mathbf{p}_i\}$ are used rather than velocities for reasons of computational economy.

In this scheme, the change in momentum of particle i at each time step is determined by the total force on that particle. Thus, one is free to choose how to partition the total force between the short range and the smoothly varying part. The reference force $\mathbf{F}$ is the interparticle force that the mesh calculation represents. For reasons of optimization, the cutoff radius of $\mathbf{F}_{sr}$ has to be as small as possible, and therefore $\mathbf{F}$ to be equal to the total interparticle force down to as small a particle separation as possible. However, this is not possible due to the limited memory storage and the required CPU time even in the state-of-the-art computers.

The harmonic content of the reference force is reduced by smoothing. A suitable form of reference force for a Coulombic long-range force is one which follows the point particle force law beyond the cutoff radius r_e, and goes smoothly to zero within that radius. The smoother the decay of $\mathbf{F}(\mathbf{x})$ and the large $\mathbf{r}_e$ becomes, the more rapidly the harmonics $\mathbf{R}(\mathbf{k})$ decay with increasing $\mathbf{k}$. Such smoothing procedure is equivalent to ascribing a finite size to the charged particle. As a result, a straightforward method of including smoothing is to ascribe some simple density profile $S(\mathbf{x})$ to the reference interparticle force. Examples of shapes, which are used in practice, and give comparable total force accuracy are the uniformly charged sphere, the sphere with uniformly decreasing density, of the form

$$S(\mathbf{r}) = \begin{cases} \dfrac{48}{\pi a^4}\left(\dfrac{a}{2} - r\right), & r < a/2 \\ 0, & \text{otherwise} \end{cases} \tag{6.29}$$

and the Gaussian distribution of density. The second scheme gives marginally better accuracies in 3D simulations. Note that the cutoff radius of the short-range force implied by Eq. (6.29) is a rather than $\mathbf{r}_e$. In practice, one can make r significantly smaller than a, because continuity of the derivatives at $r = a$ causes the reference force to closely follow the point particle force for radii somewhat less than a. It has been found empirically that a good measure of the lower bound of $\mathbf{r}_e$ is given by the cube root of the autocorrelation volume of the charge shapes, which for the case of uniformly decreasing density gives

$$r_e \geq \left(\frac{5\pi}{48}\right)^{1/3} a \approx 0.7a. \tag{6.30}$$

Once the reference interparticle force $\mathbf{F}$ for the PM part of the calculation is chosen, the short-range part $\mathbf{F}_{sr}$ is found by subtracting $\mathbf{F}$ from the total interparticle force, i.e.,

$$\mathbf{F}_{sr} = \mathbf{F}^{tot} - \mathbf{F}. \tag{6.31}$$

6.5.2.2 The Corrected Coulomb Approach

This second approach is a purely numerical scheme that generates a corrected Coulomb force look-up table for the individual e–e and e–i interaction terms. To calculate the proper short-range force, one has to define a 3D box with uniform mesh spacing in each direction. A single (fixed) electron is then placed at a known position within a 3D domain, while a second (target) electron is swept along the "device" in, for example, 0.2 nm increments so that it passes through the fixed electron. The 3D box is usually made sufficiently large so that the boundary conditions do not influence the potential solution. The electron charges are assigned to the nodes using one of the charge-assignment schemes discussed previously [114]. A 3D Poisson equation solver is then used to solve for the node or mesh potentials. At self-consistency, the force on the swept electron $\mathbf{F} = \mathbf{F}_{mesh}$ is interpolated from the mesh or node potential. In a separate experiment, the Coulomb force $\mathbf{F}_{coul} = \mathbf{F}_{tot}$ is calculated using standard Coulomb law. For each electron separation, one then tabulates $\mathbf{F}_{mesh}$, $\mathbf{F}_{coul}$ and the difference between the two $\mathbf{F}' = \mathbf{F}_{coul} - \mathbf{F}_{mesh} = \mathbf{F}_{sr}$, which is called the corrected Coulomb force or a short-range force. The later is stored in a separate look-up table.

As an example, the corresponding fields to these three forces for a simulation experiment with mesh spacing of 10 nm in each direction are shown in Figure 6.17. It is clear that the mesh force and the Coulomb force are identical when the two electrons are separated several mesh points (30–50 nm apart). Therefore, adding the two forces in this region would result in double-counting of the force. Within 3–5 mesh points, $\mathbf{F}_{mesh}$ starts to deviate from $\mathbf{F}_{coul}$. When the electrons are within the same mesh cell, the mesh force approaches zero, due to the

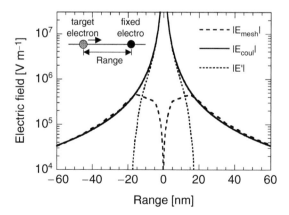

FIGURE 6.17: Mesh, Coulomb and corrected Coulomb field versus the distance between the two electrons

Note: $F = -eE$

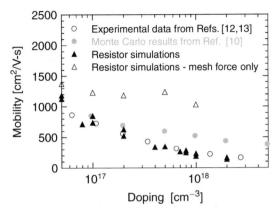

FIGURE 6.18: Low-field electron mobility derived from 3D resistor simulations versus doping. Also shown on this figure are the Ensemble Monte Carlo results and the appropriate experimental data

smoothing of the electron charge when divided amongst the nearest node points. The generated look-up table for $\mathbf{F}'$ also provides important information concerning the determination of the minimum cutoff range based upon the point where $\mathbf{F}_{coul}$ and $\mathbf{F}_{mesh}$ begin to intersect, i.e., $\mathbf{F}'$ goes to zero.

Figure 6.18 shows the simulated doping dependence of the low-field mobility, derived from 3D resistor simulations, which is a clear example demonstrating the importance of the proper inclusion of the short-range electron–ion interactions. For comparison, also shown in this figure are the simulated mobility results reported in [127], calculated with a bulk EMC technique using the Brooks–Herring approach [128] for the e–i interaction, and finally the measured data [129] for the case when the applied electric field is parallel to the ⟨100⟩ crystallographic direction. From the results shown, it is clear that adding the corrected Coulomb force to the mesh force leads to mobility values that are in very good agreement with the experimental data. It is also important to note that, if only the mesh force is used in the free-flight portion of the simulator, the simulation mobility data points are significantly higher than the experimental ones due to the omission of the short-range portion of the force.

The short-range e–e and e–i interactions also play significant role in the operation of semiconductor devices. For example, carrier thermalization at the drain end of the MOSFET channel is significantly affected by the short-range e–e and e–i interactions. This is illustrated in Figure 6.19 for the example of a 80 nm channel-length n-MOSFET. Carrier thermalization occurs over distances that are on the order of few nm when the e–e and e–i interactions are included in the problem. Using the mesh force alone does not lead to complete thermalization of the carriers along the whole length of the drain extension, and this can lead to inaccuracies when estimating the device on-state current.

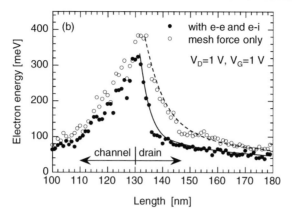

FIGURE 6.19: Average energy of the electrons coming to the drain from the channel. Filled (open) circles correspond to the case when the short-range e–e and e–i interactions are included (omitted) in the simulations. The channel length extends from 50 to 130 nm

PROBLEMS FOR CHAPTER 6:

1. In the so-called rigid ion approximation, the ions of the crystal lattice are assumed to execute only small oscillations around their equilibrium positions and it is also assumed that the ions vibrate rigidly, carrying their potentials with them as they move. With the above approximations, the total potential energy due to the electron–ion interaction can be expanded in a Taylor series, i.e.,

$$V(\mathbf{r}) = \sum_i V(\mathbf{r} - \mathbf{R}_i) = \sum_i V(\mathbf{r} - \mathbf{R}_{i0}) + \sum_i \nabla V(\mathbf{r} - \mathbf{R}_i)\big|_0 \cdot \mathbf{u}_i + \dots$$
$$\approx \sum_i V(\mathbf{r} - \mathbf{R}_{i0}) + H_{\text{ep}}$$

(1)

In (1), $\mathbf{r}$ is the electron coordinate, $\mathbf{R}_{i0}$ ($\mathbf{R}_i$) is the equilibrium (actual) position of the ith ion, $\mathbf{u}_i$ is the ion displacement and H_{ep} is the perturbing potential that involves only the first-order term in the Taylor series expansion. Calculate the matrix element for scattering between some initial state Ψ_i and some final state Ψ_f. For the one-electron states use plane waves instead of Bloch functions. For the potential $V(\mathbf{r})$ that appears in (1), assume to be the simple Coulomb potential of the form

$$V(\mathbf{r}) = -\frac{Ze^2}{4\pi\varepsilon_0 \mathbf{r}},$$

where Ze is the charge of the ion. Comment on the behavior of the matrix element for small $\mathbf{q}$ ($\mathbf{q}$ being the momentum transfer in the scattering process). Consider normal processes only. What is wrong with this simple model?

2. Assuming nonparabolic dispersion relation for the electrons, $\hbar^2 k^2 / 2m^* = E(1 + \alpha E)$, where α is the nonparabolicity factor, evaluate the scattering rate out of state $\mathbf{k}$ for intravalley nonpolar optical phonon scattering. Verify your answer with the result presented in class using $\alpha = 0$.

3. For alloys of compound semiconductors, such as $Al_x Ga_{1-x} As$, microscopic fluctuations in the alloy composition x produce perturbations in the conduction and valence band edges. The transition rate for alloy scattering is given by

$$S(k, k') = \frac{2\pi}{\hbar} \left(\frac{3\pi^2}{16} \right) \frac{|\Delta U|^2}{N\Omega} \delta(E' - E),$$

where N is concentration of atoms and

$$\Delta U = x(1 - x)(\chi_{GaAs} - \chi_{AlAs}),$$

where χ is the electron affinity.

- Explain why alloy scattering vanishes at $x = 0$ and $x = 1$.
- Derive an expression for the momentum relaxation time $\tau_m(\mathbf{k})$ for alloy scattering.

4. Using the instructions given in class, create the scattering table for GaAs bulk Monte Carlo that incorporates the following scattering mechanisms:

- polar optical phonon scattering for the Γ, L, and X valleys
- Intervalley scattering between the Γ, L, and X valleys.

Given below are all the parameters that you need in your model:

0.063   ! Relative mass for gamma-valley (rel_mass_gamma)
0.170   ! Relative mass for the L-valley (rel_mass_L)
0.58   ! Relative mass for the X-valley (rel_mass_X)

10.92   ! High-frequency relative permittivity (eps_high)
12.9   ! Low-frequency relative permittivity (eps_low)

0.62   ! Nonparabolicity factor for the gamma-valley (nonparabolicity_gamma)
0.50   ! Nonparabolicity factor for the L-valley (nonparabolicity_L)
0.30   ! Nonparabolicity factor for the X-valley (nonparabolicity_X)

0.29   ! Potential energy difference between gamma and L valley (split_L_gamma)
0.48   ! Potential energy difference between gamma and X valleys (split_X_gamma)

1.   ! Equivalent gamma-valleys (eq_valleys_gamma)
4.   ! Equivalent L-valleys (eq_valleys_L)
3.   ! Equivalent X-valleys (eq_valleys_X)

5370.   ! Crystal density (density)
5.22E3   ! Sound velocity (sound_velocity)

The parameters for Phonon Scattering are as follows:

polar_en_gamma = 0.03536 ! [eV]
polar_en_L = 0.03536 ! [eV]
polar_en_X = 0.03536 ! [eV]

DefPot_gamma_L = 1.8E10 ! [eV/m]
DefPot_gamma_X = 10.E10 ! [eV/m]
DefPot_L_gamma = 1.8E10 ! [eV/m]
DefPot_L_L = 5.E10 ! [eV/m]
DefPot_L_X = 1.E10 ! [eV/m]
DefPot_X_gamma = 10.E10 ! [eV/m]
DefPot_X_L = 1.E10 ! [eV/m]
DefPot_X_X = 10.E10 ! [eV/m]

    phonon_gamma_L = 0.0278 ! [eV]
    phonon_gamma_X = 0.0299 ! [eV]
    phonon_L_gamma = 0.0278 ! [eV]
    phonon_L_L = 0.029 ! [eV]
    phonon_L_X = 0.0293 ! [eV]
    phonon_X_gamma = 0.0299 ! [eV]
    phonon_X_L = 0.0293 ! [eV]
    phonon_X_X = 0.0299 ! [eV]

Assume that the maximum energy in your scattering table is 2 eV. Plot the cumulative scattering tables for the Γ , L, and X valleys separately.

5. Derive and expression for the polar angle after scattering for polar optical phonon scattering for nonparabolic bands.

6. As discussed in class, updating the carrier momentum after scattering is most easily accomplished in the rotated coordinate system. The rotated x-axis is related to the original x-axis by $\hat{x}_r = Y_\theta Z_\phi \hat{x}$, where Y_θ describes a rotation of θ about the y-axis, and

Z_ϕ describes a rotation of ϕ about the z-axis. The angles θ and ϕ represent the polar and the azimuthal angles of the carrier momentum in the original coordinate system before the scattering event. Calculate the rotation matrices Y_θ and Z_ϕ. If the azimuthal and the polar angles after the scattering event in the rotated coordinate system are α and β, and the scattering process is elastic, calculate the wavevector components along the principal axes in the original coordinate system after the scattering event.

7. Develop an Ensemble Monte Carlo code for electrons for bulk GaAs material system. Use a three-band parabolic band model, and include acoustic, polar optical and intervalley scattering in your theoretical model. For the evaluation of the scattering rates use the values given in Problem 4 modified for the nonparabolicity of the bands. Use $T = 300$ K. Assume that the uniform electric field that is applied is along the z-direction. Provide the following outputs:

- Plot the histograms for the initial carrier energy and the z-component of the carrier wavevector.

- Plot the time evolution of the average electron velocity, average electron energy and the valley population for uniform electric fields equal to 0.5, 1.0, 2.0, and 5.0 kV cm^{-1}. Explain the origin for the velocity overshoot in this material system for high electric fields. For electric field of 2 kV cm^{-1}, also plot the time evolution of the x- and y-components of the electron velocity.

- Plot the steady-state results for the average drift velocity, average electron energy and the valley population versus electric field. Vary the electric field in the range from 0.1 to 10 kV cm^{-1}. What is the value of the low-field electron mobility for bulk GaAs material? Also, explain what is the origin for the average drift velocity decrease at high electric fields.

In your steady-state calculations, make sure that all transients have died away when computing the time-averaged quantities of interest.

8. Develop a 2D Multigrid solver for a prototypical MESFET device. Follow the instructions given in appendix A.

9. Couple your 2D Multigrid Poisson equation solver with your bulk Monte Carlo solver to get an Ensemble Monte Carlo device simulator for an n-type MESFET with the following structure: active-layer thickness of 0.12 μm and doping of 10^{17} cm^{-3}; source/drain contact length $= 1.0$ μm; gate to source/drain spacing $= 1.35$ μm; and gate metal work function $= 4.77$ eV. The way you combine the two is described in the Xiaojiang He thesis that is posted on the web-site (www.eas.asu.edu/~vasilesk). Briefly, you have to follow this procedure:

- For given device geometry first solve for the electron density using equilibrium statistics.

- Given the electron density values, calculate how many electrons you need to initialize in each cell of the device.

- Apply bias on the structure and let the time run to 10 ps.

- At every 0.2 fs update the electric fields.

- Register the charges entering and exiting a contact to calculate the current.
  For the charge assignment scheme use the nearest element cell method (NEC scheme—see paper by Laux). Having made your device simulator operational, perform the following:

  - For gate lengths of 0.3, 1.0, and 1.5 μm, plot the ID–VDS curves for VGS = 0. Save the results for VDS = 0.5 V and for VDS = 3 V. Make a 2D plot of the electron density and explain what does the results for VDS = 3 V imply about the current saturation mechanism. At both bias points, plot the electron densities corresponding to the electrons residing in the gamma, X and L valley separately.

  - For the case of gate length of 0.3 μm, plot ID vs. VG (linear and log) for VDS = 3 V. Change the gate to source/drain spacing from 1.35 to 0.3 μm. Replot the ID vs. VG curves. Compare the results.

Numerical Solution of Algebraic Equations

The solution of linear systems of algebraic equations is an important subject of linear algebra [130], and the computational considerations needed for computer implementation are usually treated in some detail in introductory numerical methods courses. This section simply represents a quick review or overview of the subject—it is not intended as a complete treatise on this topic. Students with little or no background in this area are referred to one of many good numerical methods texts that treat the subject in more detail. The numerical solution of large systems of algebraic equations is a direct consequence of the finite–difference method for solving ordinary differential equations (ODEs) or partial differential equations (PDEs). Recall that the goal in these techniques is to break the continuous differential equation into a coupled set of algebraic difference equations for each finite volume or node in the system. When one has only a single independent variable (the ODE case), this process can easily lead to several hundred simultaneous equations that need to be solved. For multiple independent variables (the PDE case), systems with hundreds of thousands of equations are common. Thus, in general, we need to be able to solve large systems of linear equations of the form $Ax = b$ as part of the solution algorithm for general finite–difference methods.

There are two general schemes for solving linear systems: direct elimination methods and iterative methods. All the direct methods are, in some sense, based on the standard Gauss elimination technique, which systematically applies row operations to transform the original system of equations into a form that is easier to solve. In particular, this section overviews an algorithm for implementation of the basic Gauss elimination scheme and it also highlights the LU Decomposition method which, although functionally equivalent to the Gauss elimination method, does provide some additional flexibility for computer implementation. Thus, the LU decomposition method is often the preferred direct solution method for low to medium sized systems (usually less than 200–300 equations).

For large systems, iterative methods (instead of direct elimination methods) are almost always used. This switch is required from accuracy considerations (related to round-off errors), from memory limitations for physical storage of the equation constants, from considerations for treating nonlinear problems, and from overall efficiency concerns. There are several specific iterative schemes that are in common use, but most methods build upon the base Gauss–Seidel method, usually with some acceleration scheme to help convergence. Thus, our focus in this brief overview is on the basic Gauss–Seidel scheme and on the use of successive over relaxation (SOR) to help accelerate convergence. We also give a brief introduction to the incomplete-lower-upper (ILU) decomposition method and a short tutorial to the multigrid method for solving 2D and 3D problems.

A.1   DIRECT METHODS

A.1.1   Gauss Elimination Method

The Gauss elimination method forms the basis for all elimination techniques. The basic idea is to modify the original equations, using legal row operations, to give a simpler form for actual solution. The basic algorithm can be broken into two stages:

1.   Forward elimination (put equations in upper triangular form).
2.   Back substitution (solve for unknown solution vector).

To see how this works, consider the following system of equations:

$$
\begin{aligned}
a_{11}x_1 + a_{12}x_2 + \cdots a_{1N}x_N &= b_1, \\
a_{21}x_1 + a_{22}x_2 + \cdots a_{2N}x_N &= b_2, \\
&\vdots \\
a_{N1}x_1 + a_{N2}x_2 + \cdots a_{NN}x_N &= b_N.
\end{aligned}
\tag{A.1}
$$

Now, with reference to this system of N equations and N unknowns, the forward elimination step (with partial pivoting) becomes:

Step 0:   Create an augmented matrix, $\tilde{A} = [Ab]$.

Step 1:   Determine the coefficient in the ith column with the largest absolute value and interchange rows such that this element is the pivot element ($i = 1, 2, 3,$ to $N - 1$).

Step 2:   Normalize the pivot equation (i.e., divide by the i,i element).

Step 3:   Multiply normalized equation i by the j, i element of equation j.

Step 4:   Subtract the resultant equation in Step 3 from equation j.
   Repeat Steps 3 and 4 for $j = i + 1$ to N.
   Go to Step 1 for next $i = i + 1$ to $N - 1$.

and the Back Substitution Step is given by

Step 5: $\quad x_N = b'_N / a'_{NN}$

Step 6: $\quad x_i = \left(b'_i - \sum_{j=i+1}^{N} a'_{ij} x_j \right) \bigg/ a'_{ii}$

$\qquad$ repeat for $i = N - 1, \, N - 2,$ to 1

Note: The primes here indicate that the coefficients at this stage are different from the original coefficients.

A.1.2 The LU Decomposition Method

The Gauss elimination method has the disadvantage that all right-hand sides (i.e., all the b vectors of interest for a given problem) must be known in advance for the elimination step to proceed. The LU decomposition method outlined here has the property that the matrix modification (or decomposition) step can be performed independent of the right-hand side vector. This feature is quite useful in practice—therefore, the LU decomposition method is usually the direct scheme of choice in most applications.

To develop the basic method, let us break the coefficient matrix into a product of two matrices,

$$A = LU, \tag{A.2}$$

where L is a lower triangular matrix and U is an upper triangular matrix. Now, the original system of equations $Ax = b$, becomes

$$LUx = b. \tag{A.3}$$

This expression can be broken into two problems,

$$Ly = b, \quad Ux = y. \tag{A.4}$$

The rationale behind this approach is that the two systems given in Eq. (A.4) are both easy to solve; one by forward substitution and the other by back substitution. In particular, because L is a lower triangular matrix, the expression $Ly = b$ can be solved with a simple forward substitution step. Similarly, since U has upper triangular form, $Ux = y$ can be evaluated with a simple back substitution algorithm.

Thus, the key to this method is the ability to find two matrices L and U that satisfy Eq. (A.4). Doing this is referred to as the decomposition step and there are a variety of algorithms available. Three specific approaches are as follows:

- Doolittle decomposition:

$$
\begin{bmatrix}
1 & 0 & 0 & 0 \\
l_{21} & 1 & 0 & 0 \\
l_{31} & l_{32} & 1 & 0 \\
l_{41} & l_{42} & l_{43} & 1
\end{bmatrix}
\begin{bmatrix}
u_{11} & u_{12} & u_{13} & u_{14} \\
0 & u_{22} & u_{23} & u_{24} \\
0 & 0 & u_{33} & u_{34} \\
0 & 0 & 0 & u_{44}
\end{bmatrix}
=
\begin{bmatrix}
a_{11} & a_{12} & a_{13} & a_{14} \\
a_{21} & a_{22} & a_{23} & a_{24} \\
a_{31} & a_{32} & a_{33} & a_{34} \\
a_{41} & a_{42} & a_{43} & a_{44}
\end{bmatrix} . \quad (A.5)
$$

Because of the specific structure of the matrices, a systematic set of formulae for the components of L and U results.

- Crout decomposition:

$$
\begin{bmatrix}
l_{11} & 0 & 0 & 0 \\
l_{21} & l_{22} & 0 & 0 \\
l_{31} & l_{32} & l_{33} & 0 \\
l_{41} & l_{42} & l_{43} & l_{44}
\end{bmatrix}
\begin{bmatrix}
1 & u_{12} & u_{13} & u_{14} \\
0 & 1 & u_{23} & u_{24} \\
0 & 0 & 1 & u_{34} \\
0 & 0 & 0 & 1
\end{bmatrix}
=
\begin{bmatrix}
a_{11} & a_{12} & a_{13} & a_{14} \\
a_{21} & a_{22} & a_{23} & a_{24} \\
a_{31} & a_{32} & a_{33} & a_{34} \\
a_{41} & a_{42} & a_{43} & a_{44}
\end{bmatrix} . \quad (A.6)
$$

The evaluation of the components of L and U is done in a similar fashion as above.

- Cholesky factorization:
  For symmetric, positive definite matrices, where $A = A^T$ and $x^T A x > 0$ for all $x \neq 0$ then,

$$
U = L^T \quad \text{and} \quad A = LL^T \quad (A.7)
$$

and a simple set of expressions for the elements of L can be obtained (as above). Once the elements of L and U are available (usually stored in a single $N \times N$ matrix), Matlab's standard equation solver (using the backslash notation, $x = A \backslash b$), uses several variants of the basic LU Decomposition method depending on the form of the original coefficient matrix (see the Matlab help files for details).

A.1.3 LU Decomposition in 1D

The LU decomposition method is very trivial for 1D problems where the discretization of the ODE or the PDE leads to a three-point stencil and a tridiagonal matrix A. It is easy to show, that the system of equations that we need to solve, and for the purpose of clarity, we denote by

$Ax = f$, in matrix form reads

$$\begin{bmatrix} a_1 & c_1 & 0 & 0 & \cdots & 0 \\ b_2 & a_2 & c_2 & 0 & \cdots & 0 \\ 0 & b_3 & a_3 & c_3 & \cdots & 0 \\ \vdots & & & & & \\ \vdots & & & & & \\ 0 & 0 & 0 & \cdots & b_n & a_n \end{bmatrix} \begin{bmatrix} x_1 \\ x_2 \\ x_3 \\ \vdots \\ \vdots \\ x_n \end{bmatrix} = \begin{bmatrix} f_1 \\ f_2 \\ f_3 \\ \vdots \\ \vdots \\ f_n \end{bmatrix}. \tag{A.8}$$

The solution of this problem can be represented as a two-step procedure that is explained below.

Step 1: Decompose the coefficient matrix A into a product of lower and upper triangular matrices:

$$A = LU = \begin{bmatrix} 1 & 0 & 0 & 0 & \cdots & 0 \\ \beta_2 & 1 & 0 & 0 & \cdots & 0 \\ 0 & \beta_3 & 1 & 0 & \cdots & 0 \\ \vdots & & & & & \\ \vdots & & & & & \\ 0 & 0 & 0 & \cdots & \beta_n & 1 \end{bmatrix} \begin{bmatrix} \alpha_1 & c_1 & 0 & 0 & \cdots & 0 \\ 0 & \alpha_2 & c_2 & 0 & \cdots & 0 \\ 0 & 0 & \alpha_3 & c_3 & \cdots & 0 \\ \vdots & & & & & \\ \vdots & & & & & \\ 0 & 0 & 0 & \cdots & 0 & \alpha_n \end{bmatrix}. \tag{A.9}$$

From the equality of the two matrices, we have:

$$\alpha_1 = a_1; \quad \beta_k = b_k/\alpha_{k-1}; \quad \alpha_k = a_k - \beta_k c_{k-1}; \quad k = 2, 3, \ldots, n. \tag{A.10}$$

Step 2: Solve the system of equations $LUx = f$, by first solving $Lg = f$ using forward substitution, and then solving $Ux = g$ using backward substitution. Then, the solution of $Lg = f$ is represented as:

$$g_1 = f_1; \quad g_k = f_k - \beta_k g_{k-1}; \quad k = 2, 3, \ldots, n. \tag{A.11}$$

And the solution of $Ux = g$ as:

$$x_n = g_n/\alpha_n; \quad x_k = [g_k - c_k x_{k+1}]/\alpha_k; \quad k = n-1, n-2, \ldots, 2, 1 \tag{A.12}$$

A.2   ITERATIVE METHODS

For large systems of equations, an iterative solution scheme for the unknown vector can always be written in the form

$$x^{p+1} = Bx^p + c, \tag{A.13}$$

where B is the iteration matrix, c is a constant vector and p is an iteration counter. Convergence of this scheme is guaranteed if the largest eigenvalue of the iteration matrix is less that unity, where ρ = spectral radius = $|\lambda|_{\max}$. Therefore, if $\rho < 1$ the iterative scheme will converge. If $\rho \ll 1$, the iterative scheme converges very rapidly. If $\rho \approx 1$ but less than unity, the scheme will be slowly converging. The iteration algorithm will diverge if the spectral radius is greater than unity. Convergence is tested during the iterative process by computing the largest relative change from one iteration to the next, and comparing the absolute value of this result with some desired tolerance. If the maximum relative change is less than the desired accuracy, then the process is terminated. If this condition is not satisfied, then another iteration is performed.

A.2.1   The Gauss–Seidel Method

Let us take the original system of equations given by $Ax = b$ and convert it into the classical Gauss–Seidel iterative scheme. To do this, let us break the original matrix into three specific components, or

$$A = L + D + U, \tag{A.14}$$

where the three matrices on the right-hand side, in sequence, are strictly lower triangular, diagonal, and strictly upper triangular matrices. Now, substituting this into the original expression gives

$$(L + D)x + Ux = b \tag{A.15}$$

or

$$(L + D)x = b - Ux. \tag{A.16}$$

If we premultiply by $(L + D)^{-1}$ and notice that the solution vector appears on both sides of the equation, we can write the equation in an iterative form as

$$x^{p+1} = -(L + D)^{-1}Ux^p + (L + D)^{-1}b. \tag{A.17}$$

Clearly this is in the standard form for iterative solutions as defined in Eq. (A.13), where the iteration matrix is given by

$$B = -(L + D)^{-1}U \tag{A.18}$$

and the constant vector is written as

$$c = (L + D)^{-1}b. \tag{A.19}$$

This form of the iteration strategy is useful for the study of the convergence properties of model problems. It is, however, not particularly useful as a program algorithm for code implementation.

For actual implementation on the computer, one writes these equations differently, never having to formally take the inverse as indicated above. In practice, Eq. (A.17) is written in iterative form as

$$Dx^{p+1} = b - Lx^{p+1} - Ux^p \tag{A.20}$$

or

$$x^{p+1} = D^{-1}\left(b - Lx^{p+1} - Ux^p\right). \tag{A.21}$$

This specific form is somewhat odd at first glance, since x^{p+1} appears on both sides of the equation. This is justified because of the special form of the strictly lower triangular matrix, L. This can be seen more clearly if the matrix equations are written using discrete notation. In discrete form Eq. (A.21) can be expanded as

$$x_i^{p+1} = \frac{1}{a_{ii}}\left(b_i - \sum_{j=1}^{i-1} a_{ij}x_j^{p+1} - \sum_{j=i+1}^{N} a_{ij}x_j^{p}\right), \tag{A.22}$$

where the diagonal elements of D^{-1} are simply $1/a_{ii}$ and the limits associated with the summations account for the special structure of the L and U matrices.

A.2.2   The Successive Over-Relaxation (SOR) Method

To improve the rate of convergence, one might consider using a weighted average of the results of the two most recent estimates to obtain the next best guess of the solution. If the solution is converging, this might help extrapolate to the real solution more quickly. This idea is the basis of the SOR method. In particular, let α be some weight factor with a value between 0 and 2. Now, let us compute the next value of x^{p+1} to use in the Gauss–Seidel method as a linear combination of the current value, x^{p+1}, and the previous solution, x^p, as follows:

$$x^{p+1}|_{\text{new}} = \alpha x^{p+1} + (1 - \alpha)x^p \quad \text{with} \quad 0 < \alpha < 2. \tag{A.23}$$

Note that if α is unity, we simply get the standard Gauss–Seidel method (or whatever base iterative scheme is in use). When α is greater that unity, the system is said to be over-relaxed, indicating that the latest value, x^{p+1}, is being weighted more heavily (weight for x^p is negative). If, however, α is less than one, the system is under-relaxed, this time indicating that the previous solution, x^p, is more heavily weighted (positive weight values). The idea, of course, is to choose the relaxation parameter to improve convergence (reduce the spectral radius). This is most often done in a trial-and-error fashion for certain classes of problems (experience helps here). Some more advanced codes do try to estimate this quantity as part of the iterative calculation, although this is not particularly easy.

A.2.3   Other Iterative Methods

A variety of schemes for improving convergence have been developed over the years, with many taking advantage of the particular structure of the algebraic equations or some characteristic of the physical system under study. Other Iterative methods of interest include:

(a) Incomplete LU decomposition for 2D and 3D problems

Within incomplete factorization schemes [131] for 2D problems, the matrix A is decomposed into a product of lower (L) and upper (U) triangular matrices, each of which has four nonzero diagonals in the same locations as the ones of the original matrix A. The unknown elements of the L and U matrices are selected in such a way that the five diagonals common to both A and $A' = LU$ are identical and the four superfluous diagonals represent the matrix N, i.e., $A' = A + N$. Thus, rather than solving the original system of equations $Ax = b$, one solves the modified system $LUx = b + Nx$, by solving successively the matrix equations $LV = b + Nx$ and $V = Ux$, where V is an auxiliary vector. It is important to note that the four superfluous terms of N affect the rate of convergence of the ILU method. Stone [132] suggested the introduction of partial cancellation, which minimizes the influence of these additional terms and accelerates the rate of convergence of the ILU method. By using a Taylor series expansion, the superfluous terms appearing in A' are partially balanced by subtracting approximately equal terms.

(b) Multigrid method

The multigrid method represents an improvement over the SOR and ILU methods in terms of iterative techniques available for solving large systems of equations [133]. The basic principle behind the multigrid method is to reduce different Fourier components of the error on grids with different mesh sizes. Most iterative techniques work by quickly eliminating the high-frequency Fourier components, while the low-frequency ones are left virtually unchanged. The result is a convergence rate that is initially fast, but slows down dramatically as the high-frequency components disappear. The multigrid method utilizes several grids, each with consecutively coarser mesh sizes. Each of these grids acts to reduce a different Fourier component of the error, therefore increasing the rate of convergence with respect to single grid-based methods, such as an SOR.

Practical multigrid methods were first introduced in the 1970s by Brandt [134]. These methods can solve elliptic PDEs discretized on N grid points in $O(N)$ operations. The "rapid" direct elliptic solvers discussed in Ref. [135] solve special kinds of elliptic equations in $O(N \log N)$ operations. The numerical coefficients in these estimates are such that multigrid methods are comparable to the rapid methods in execution speed. Unlike the rapid methods, however, the multigrid methods can solve general elliptic equations with nonconstant coeffi-cients with hardly any loss in efficiency. Even nonlinear equations can be solved with comparable speed. Unfortunately there is not a single multigrid algorithm that solves all elliptic problems. Rather there is a multigrid technique that provides the framework for solving these problems. You have to adjust the various components of the algorithm within this framework to solve your specific problem. We can only give a brief introduction to the subject here. In this approach, the method obtains successive solutions on finer and finer grids. You can stop the solution either at a prespecified fineness, or you can monitor the truncation error due to the discretization, quitting only when it is tolerably small.

(i) From One-Grid, through Two-Grid, to Multigrid

The key idea of the multigrid method can be understood by considering the simplest case of a two-grid method. Suppose we are trying to solve the linear elliptic problem

$$Lu = f, \qquad (A.24)$$

where L is some linear elliptic operator and f is the source term. When one discretizes Eq. (A.24) on a uniform grid with mesh size h, the resulting set of linear algebraic equations arises

$$L_h u_h = f_h. \qquad (A.25)$$

Let $\tilde{u}_h$ denote some approximate solution to Eq. (A.25). We will use the symbol u_h to denote the exact solution to the difference equations. Then the *error* in $\tilde{u}_h$ or the *correction* is

$$v_h = u_h - \tilde{u}_h. \qquad (A.26)$$

The *residual* or *defect* is

$$d_h = L_h \tilde{u}_h - f_h. \qquad (A.27)$$

Since L_h is linear, the error satisfies

$$L_h v_h = -d_h. \qquad (A.28)$$

At this point we need to make an approximation to L_h in order to find v_h. The clas-sical iteration methods, such as Jacobi or Gauss–Seidel, do this by finding, at each stage, an approximate solution of the equation

$$\hat{L}_h \hat{v}_h = -d_h, \qquad (A.29)$$

where $\hat{L}_h$ is a "simpler" operator than L_h. For example, $\hat{L}_h$ is the diagonal part of L_h for Jacobi iteration, or the lower triangle for Gauss–Seidel iteration. The next approximation is generated by

$$\tilde{u}_h^{new} = \tilde{u}_h + \hat{v}_h. \tag{A.30}$$

Now consider, as an alternative, a completely different type of approximation for L_h, one in which we "coarsify" rather than "simplify." That is, we form some appropriate approximation L_H of L_h on a coarser grid with mesh size H (we will always take $H = 2h$, but other choices are possible). The residual equation is now approximated by

$$L_H v_H = -d_H. \tag{A.31}$$

Since L_H has smaller dimension, this equation will be easier to solve than Eq. (A.28). To define the defect d_H on the coarse grid, we need a *restriction operator* R that restricts d_h to the coarse grid:

$$d_H = R d_h. \tag{A.32}$$

The restriction operator is also called the *fine-to-coarse operator* or the *injection operator*. Once we have a solution $\tilde{v}_H$ to Eq. (A.31), we need a *prolongation operator* P that prolongates or interpolates the correction to the fine grid:

$$\tilde{v}_H P \tilde{v}_H. \tag{A.33}$$

The prolongation operator is also called the *coarse-to-fine operator* or the *interpolation operator*. Both R and P are chosen to be linear operators. Finally, the approximation $\tilde{u}_h$ can be updated:

$$\tilde{u}_h^{new} = \tilde{u}_h + \tilde{v}_h \tag{A.34}$$

One step of this *coarse–grid correction scheme* is thus:

- Compute the defect on the fine grid from Eq. (A.27).
- Restrict the defect by Eq. (A.32).
- Solve Eq. (A.31) exactly on the coarse grid for the correction.
- Interpolate the correction to the fine grid by Eq. (A.33).
- Compute the next approximation by Eq. (A.34).

Let us contrast the advantages and disadvantages of relaxation and the coarse-grid correction scheme. Consider the error v_h expanded into a discrete Fourier series. Call the components in the lower half of the frequency spectrum the *smooth components* and the high-frequency components the *nonsmooth components*. We have seen that relaxation becomes very slowly convergent in the limit $h \to 0$, i.e., when there are a large number of mesh points. The reason turns out to be that the smooth components are only slightly reduced in amplitude on each iteration. However, many relaxation methods reduce the amplitude of the nonsmooth components by large factors on each iteration: They are good *smoothing operators*. For the two-grid iteration, on the other hand, components of the error with wavelengths $2H$ are not even representable on the coarse grid and so cannot be reduced to zero on this grid. But it is exactly these high-frequency components that can be reduced by relaxation on the fine grid! This leads us to combine the ideas of relaxation and coarse-grid correction:

- Presmoothing: Compute $\bar{u}_h$ by applying $\nu_1 \geq 0$ steps of a relaxation method to $\tilde{u}_h$.
- Coarse-grid correction: As above, using $\bar{u}_h$ to give $\bar{u}_h^{\text{new}}$.
- Post-smoothing: Compute $\tilde{u}_h^{\text{new}}$ by applying $\nu_2 \geq 0$ steps of the relaxation method to $\bar{u}_h^{\text{new}}$.

It is only a short step from the above two-grid method to a multigrid method. Instead of solving the coarse-grid defect Eq. (A.31) exactly, we can get an approximate solution of it by introducing an even coarser grid and using the two-grid iteration method. If the convergence factor of the two-grid method is small enough, we will need only a few steps of this iteration to get a good enough approximate solution. We denote the number of such iterations by γ. Obviously, we can apply this idea recursively down to some coarsest grid. There the solution is found easily, for example by direct matrix inversion or by iterating the relaxation scheme to convergence. One iteration of a multigrid method, from finest grid to coarser grids and back to finest grid again, is called a *cycle*. The exact structure of a cycle depends on the value of γ, the number of two-grid iterations at each intermediate stage. The case $\gamma = 1$ is called a V-cycle, while $\gamma = 2$ is called a W-cycle (see Figure A.1). These are the most important cases in practice. Note that once more than two grids are involved, the pre-smoothing steps after the first one on the finest grid need an initial approximation for the error v. This should be taken to be zero.

(ii) Smoothing, Restriction, and Prolongation Operators

The most popular smoothing method, and the one you should try first, is Gauss–Seidel, since it usually leads to a good convergence rate. The exact form of the Gauss–Seidel method depends on the ordering chosen for the mesh points. For typical second-order elliptic equations like our model problem, it is usually best to use red-black ordering, making one pass through the mesh updating the "even" points (like the red squares of a checkerboard) and another pass updating the "odd" points (the black squares). When quantities are more strongly coupled along one dimension than another, one should relax a whole line along that dimension simultaneously. Line relaxation for nearest-neighbor coupling involves solving a tridiagonal system, and so is still efficient. Relaxing odd and even lines on successive passes is called zebra relaxation and is

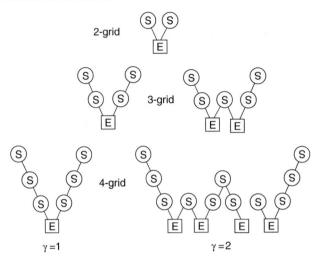

FIGURE A.1: Structure of multigrid cycles. S denotes smoothing, while E denotes exact solution on the coarsest grid. Each descending line\denotes restriction® and each ascending line/denotes prolongation (P). The finest grid is at the top level of each diagram. For the V-cycles ($\gamma = 1$) the E step is replaced by one 2-grid iteration each time the number of grid levels increases by one. For the W-cycles ($\gamma = 2$) each E step gets replaced by two 2-grid iterations

usually preferred over simple line relaxation. Note that SOR should not be used as a smoothing operator. The over-relaxation destroys the high-frequency smoothing that is so crucial for the multigrid method.

A succinct notation for the prolongation and restriction operators is to give their symbol. The symbol of P is found by considering v_H to be 1 at some mesh point (x, y), zero elsewhere, and then asking for the values of Pv_H. The most popular prolongation operator is simple bilinear interpolation. It gives nonzero values at the nine points (x, y), $(x + h, y)$, $\ldots$, $(x - h, y - h)$, where the values are $1, \frac{1}{2}, \ldots, \frac{1}{4}$.

Its symbol is therefore

$$
\begin{bmatrix}
\frac{1}{4} & \frac{1}{2} & \frac{1}{4} \\
\frac{1}{2} & 1 & \frac{1}{2} \\
\frac{1}{4} & \frac{1}{2} & \frac{1}{4}
\end{bmatrix}.
\tag{A.35}
$$

The symbol of R is defined by considering v_h to be defined everywhere on the fine grid, and then asking what is Rv_h at (x, y) as a linear combination of these values. The simplest possible choice for R is *straight injection*, which means simply filling each coarse-grid point with the value from the corresponding fine-grid point. Its symbol is "[1]." However, difficulties can arise in practice with this choice. It turns out that a safe choice for R is to make it the adjoint operator to P. Then, take P to be bilinear interpolation, and choose $u_H = 1$ at (x, y), zero elsewhere. Then take P to be bilinear interpolation, and choose $u_H = 1$ at (x, y), zero elsewhere. Finally, the symbol of R is

$$\begin{bmatrix} \frac{1}{16} & \frac{1}{8} & \frac{1}{16} \\ \frac{1}{8} & \frac{1}{4} & \frac{1}{8} \\ \frac{1}{16} & \frac{1}{8} & \frac{1}{16} \end{bmatrix}. \tag{A.36}$$

Note the simple rule: The symbol of R is 1/4 the transpose of the matrix defining the symbol of P. This rule is general whenever $R = P^\dagger$ and $H = 2h$. The particular choice of R in Eq. (A.36) is called *full weighting*. Another popular choice for R is *half weighting*, "halfway" between full weighting and straight injection. Its symbol is

$$\begin{bmatrix} 0 & \frac{1}{8} & 0 \\ \frac{1}{8} & \frac{1}{2} & \frac{1}{8} \\ 0 & \frac{1}{8} & 0 \end{bmatrix} \tag{A.37}$$

A similar notation can be used to describe the difference operator L_h. For example, the standard differencing of the model problem, Eq. (A.29), is represented by the *five-point difference star*

$$L_h = \frac{1}{h^2} \begin{bmatrix} 0 & 1 & 0 \\ 1 & -4 & 1 \\ 0 & 1 & 0 \end{bmatrix}. \tag{A.38}$$

If you are confronted with a new problem and you are not sure what P and R choices are likely to work well, here is a safe rule: Suppose m_p is the order of the interpolation P (i.e., it interpolates polynomials of degree $m_p - 1$ exactly). Suppose m_r is the order of R, and that R is the adjoint of some P (not necessarily the P you intend to use). Then if m is the order of the differential operator L_h, you should satisfy the inequality $m_p + m_r > m$. For example, bilinear interpolation and its adjoint, full weighting, for Poisson's equation satisfy $m_p + m_r = 4 > m = 2$.

Of course the P and R operators should enforce the boundary conditions for your problem. The easiest way to do this is to rewrite the difference equation to have homogeneous boundary conditions by modifying the source term if necessary. Enforcing homogeneous boundary conditions simply requires the P operator to produce zeros at the appropriate boundary points. The corresponding R is then found by $R = P^\dagger$.

APPENDIX B

Mobility Modeling and Characterization

Electrons and holes are accelerated by the electric fields, but lose momentum as a result of various scattering processes. These scattering mechanisms include lattice vibrations (phonons), impurity ions, other carriers, surfaces, and other material imperfections. A detailed chart of most of the imperfections that cause the carrier to scatter in a semiconductor is given in Figure B.1.

Since the effects of all of these microscopic phenomena are lumped into the macroscopic mobilities introduced by the transport equations, these mobilities are therefore functions of the local electric field, lattice temperature, doping concentration, and so on. Mobility modeling is normally divided into: (i) low-field behavior, (ii) high-field behavior, (iii) bulk semiconductor regions, and (iv) inversion layers. The low electric field behavior has carriers almost in equilibrium with the lattice and the mobility has a characteristic low-field value that is commonly denoted by the symbol $\mu_{n0, p0}$. The value of this mobility is dependent upon phonon and impurity scattering, both of which act to decrease the low-field mobility. The high electric field behavior shows that the carrier mobility declines with electric field because the carriers that gain energy can take part in a wider range of scattering processes. The mean drift velocity no longer increases linearly with increasing electric field, but rises more slowly. Eventually, the velocity does not increase any more with increasing field, but saturates at a constant velocity. This constant velocity is commonly denoted by the symbol v_{sat}. Impurity scattering is relatively insignificant for energetic carriers, and so v_{sat} is primarily a function of the lattice temperature.

In the early days, most experimental work on inversion layer mobilities has concentrated on *Hall* and *field-effect* mobilities. However, it is the *effective* mobility which appears in all theoretical models of MOS transistors and which is, therefore, most useful in modern MOS device modeling. Of lesser importance is the so-called *saturation* mobility. The Hall mobility, described in Section B.1(a), represents the bulk mobility and the interface, as well as the quantization effect, plays a minor role in its determination. The field-effect, effective and saturation mobilities, used to characterize MOSFET's, are described in Section B.1(b). The mobility models used in prototypical device simulator are categorized in Section B.2.

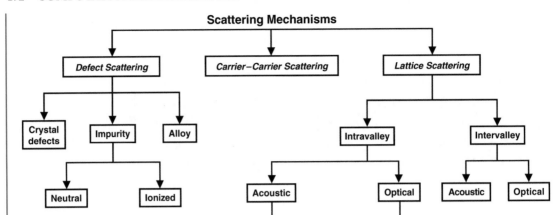

FIGURE B.1: Scattering mechanisms in a typical semiconductor

B.1   EXPERIMENTAL MOBILITIES

B.1.1   (a) Hall Mobility

The Hall measurement technique [136] is commonly used for resistivity measurements, carrier concentration characterization as well as mobility measurements. The basic setup of the Hall technique is given in Figure B.2. As shown in the figure, the applied electric field along the x-axis gives rise to a current I_x. The Lorentz force $F_y = ev_x B_z$ due to the applied magnetic field along the positive z-axis pushes the carriers upwards. This results in a pile up of electrons and holes at the top part of the sample which, in turn, gives rise to electric fields E_{yn} and E_{yp}, respectively. The transverse electric fields along the y-axis are called *Hall fields*. Since there is

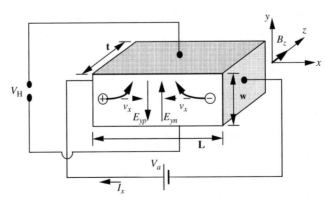

FIGURE B.2: Experimental setup for Hall measurement technique

no net current along the y-direction in steady state, the induced electric fields along the y-axis exactly balance the Lorentz force, i.e.,

$$\frac{V_H}{w} = R_H J_x B_z. \tag{B.1}$$

In (B.1), J_x is the current density and R_H is the so-called Hall coefficient. If both electrons and holes are present in the sample, the Hall coefficient is given by

$$R_H = \frac{r_h p - r_e b^2 n}{e(p + bn)^2}, \tag{B.2}$$

where $n(p)$ is the electron(hole) concentration, $b = \mu_e/\mu_h$ is the mobility ratio and r_e (r_h) is the so-called Hall scattering factor for electrons (holes) that takes into account the energy spread of the carriers. The Hall scattering factor that appears in (B.2), is defined by the ratio

$$r = \frac{\langle \tau^2 \rangle}{\langle \tau \rangle^2}, \tag{B.3}$$

where τ is the mean-free time between carrier collisions, and the average value of the mth power of τ in d-dimensions is calculated from

$$\langle \tau^m \rangle = \frac{\int_0^\infty \varepsilon^{d/2} \tau^m(\varepsilon) \, (\partial f_0/\partial \varepsilon) \, d\varepsilon}{\int_0^\infty \varepsilon^{d/2} \, (\partial f_0/\partial \varepsilon) \, d\varepsilon}, \tag{B.4}$$

where f_0 is the equilibrium Fermi-Dirac distribution function.

The Hall mobility μ_H is defined as a product of the Hall coefficient R_H and conductivity σ_x

$$\mu_H = |R_H| \sigma_x, \tag{B.5}$$

which is calculated from

$$\sigma_x = \frac{I_x L}{wt V_a}. \tag{B.6}$$

It is important to point out that the Hall mobility has to be distinguished from the so-called conductivity (or effective) mobility which does not contain the Hall scattering factor. The two mobilities are related to each other according to

$$\mu_H = r \mu_{\text{eff}}. \tag{B.7}$$

B.1.2 (b) MOSFET Mobilities

Electron mobility in surface-inversion layers has been of considerable interest for many years. At present, several mobilities (already mentioned in the introduction part of this appendix) are used to characterize MOSFETs [137].

The *effective mobility* μ_{eff} is usually deduced from the first-order one-dimensional model in the linear mode. At low drain voltages ($V_{\mathrm{DS}} = 10 - 50$ mV, where $V_{\mathrm{DS}} \ll V_{\mathrm{GS}} - V_{\mathrm{T}}$), the effective mobility is related to the *drain conductance*

$$g_{\mathrm{D}} = \left. \frac{\partial I_{\mathrm{D}}}{\partial V_{\mathrm{DS}}} \right|_{V_{\mathrm{GS}}=\mathrm{const.}}, \tag{B.8}$$

according to

$$\mu_{\mathrm{eff}} \approx \frac{L g_{\mathrm{D}}}{Z C_{\mathrm{ox}} (V_{\mathrm{GS}} - V_{\mathrm{T}})}. \tag{B.9}$$

In expressions (B.8) and (B.9), I_{D} is the drain current, L is the length and Z is the width of the channel, V_{GS} is the gate voltage, and V_{T} is the so-called threshold gate voltage. The threshold gate voltage is often defined as the voltage where the Fermi level is as close to the conduction (or valence) band at the surface as to the valence (or conduction) band in the bulk. It is experimentally determined by using various linear extrapolation techniques on the $I_{\mathrm{D}} - V_{\mathrm{GS}}$ curves, as explained in [136]. The inaccuracies in the threshold voltage significantly affect the effective mobility results. Both thermal broadening and trapping tend to obscure the accurate measurements of V_{T} and, therefore, μ_{eff}.

The previously described effective mobility is distinct from the so-called *field-effect mobility* $\mu_{\mathbf{FE}}$ which is obtained from the MOSFET *transconductance*

$$g_{\mathrm{m}} = \left. \frac{\partial I_{\mathrm{D}}}{\partial V_{\mathrm{GS}}} \right|_{V_{\mathrm{DS}}=\mathrm{const.}} \tag{B.10}$$

through the expression

$$\mu_{\mathrm{FE}} = \frac{L g_{\mathrm{m}}}{Z C_{\mathrm{ox}} V_{\mathrm{DS}}}. \tag{B.11}$$

The experimentally measured field-effect mobility is usually smaller than the effective mobility. The discrepancy between the effective and field-effect mobility is associated with the neglect of the electric-field dependence (more precisely, the neglect of the gate voltage dependence) in the derivation of the expression for μ_{FE}. For example, for the device in the linear regime and using the definitions given in (B.8) and (B.10), after a straightforward calculation it follows that the two mobilities can be related to each other according to

$$\mu_{\mathrm{FE}} \approx \mu_{\mathrm{eff}} + (V_{\mathrm{GS}} - V_{\mathrm{T}}) \left.\frac{\partial \mu_{\mathrm{eff}}}{\partial V_{\mathrm{GS}}}\right|_{V_{\mathrm{DS}}=\mathrm{const.}}. \tag{B.12}$$

Since the effective mobility decreases with the gate voltage, i.e., $\partial \mu_{\mathrm{eff}}/\partial V_{\mathrm{GS}} < 0$ (except for very low gate voltages, where it actually increases due to the decreased importance of Coulomb scattering), $\mu_{\mathrm{FE}} < \mu_{\mathrm{eff}}$. Therefore, if μ_{FE} is used for device modeling, the currents and device switching speeds are going to be underestimated.

Very rarely, the MOSFET mobility is obtained from the output current–voltage characteristics with the device in saturation. In this regime, the saturation drain current $I_{\mathrm{D,sat}}$ is calculated from

$$I_{\mathrm{D,sat}} = \frac{B Z \mu_{\mathrm{sat}} C_{\mathrm{ox}}}{2L} (V_{\mathrm{GS}} - V_{\mathrm{T}})^2, \tag{B.13}$$

where B is the body factor, which is not always well known. If one plots the variation of $\sqrt{I_{\mathrm{D,sat}}}$ vs. $(V_{\mathrm{GS}} - V_{\mathrm{T}})$, then the so-called *saturation mobility* is determined from the slope m of this curve, according to

$$\mu_{\mathrm{sat}} = \frac{2Lm^2}{B Z C_{\mathrm{ox}}}. \tag{B.14}$$

Again, due to the neglect of the gate-voltage dependence in the definition for the saturation mobility, the experimental results for μ_{sat} are always smaller compared to the ones obtained for μ_{eff}.

B.2   MOBILITY MODELING

As already noted, mobility modeling is normally divided into: (1) low- and high-field behavior, and (2) bulk semiconductor regions and inversion layers. Mobility models fall into one of the three broad categories: physically-based, semi-empirical, and empirical. Physically-based models are those that are obtained from a first-principles calculation, i.e., both the coefficients and the power dependencies appearing in the model are obtained from a fundamental calculation. In practice, physically-based models rarely agree with experimental data since considerable simplifying assumptions are made in order to arrive at a closed form solution. Therefore, to reconcile the model with experimental data, the coefficients appearing in the physically-based model are allowed to vary from their original values. In this process the power-law dependencies resulting from the first-principles calculation are preserved, and the resulting model is termed as semi-empirical. At the other end of the spectrum are empirically-based models in which the power-law dependencies are also allowed to vary. Empirical models have less physical content compared to the other two models, and also exhibit a narrower range of validity. Empirical models are usually resorted to when the dependencies predicted by the first-principles calculation do not allow a good fit between the experimental data and the corresponding semiempirical model.

At low-fields and bulk samples carriers are almost in equilibrium with the lattice vibrations and the low-field mobility is mainly affected by phonon and Coulomb scattering. At higher electric fields mobility becomes field-dependent parameter and it decreases with increasing electric field due to increased lattice scattering at higher carrier energies. In general, the bulk mobility modeling is a three-step procedure:

- Characterize low-field mobility μ_0 as a function of doping and lattice temperature T.
- Characterize the saturation velocity v_{sat} as a function of the temperature T.
- Describe the transition between the low-field and high-field regions.

Modeling carrier mobilities in inversion layers introduces additional complications. Carriers in inversion layers are subject to surface scattering, extreme carrier–carrier scattering, and quantum-mechanical size quantization effects. These effects must be accounted for in order to perform accurate simulation of MOS devices. The transverse electric field is often used as a parameter that indicates the strength of inversion layer phenomena. It is possible to define multiple nonconflicting mobility models simultaneously. It is also necessary to know which models are over-riding when conflicting models are defined.

The low-field mobility models for bulk materials include:

- constant mobility model
- Caughey and Thomas model (doping and temperature dependent mobilities) [138]
- Arora model (includes doping and temperature dependence) [139]
- Dorkel-Leturg model (includes dependence on temperature, doping and carrier-carrier scattering) [140]
- Klaassen unified low-field mobility model (provides unified description of majority and minority carrier mobility. In doing so, it includes the effects of lattice scattering, screened Coulomb charges, carrier–carrier scattering and impurity clustering effects at high concentrations) [141]

To obtain accurate results for MOSFET simulations, it is necessary to account for the mobility degradation that occurs inside inversion layers. The degradation normally occurs as a result of the substantially higher surface scattering near the semiconductor to insulator interface. This effect is handled within ATLAS by three distinct methods:

- a surface degradation model SURFMOB
- a transverse electric field model SHIRAHATA [142]
- specific inversion layer mobility models CVT (Lombardi) [143], YAMAGUCHI [144] and TASCH [145]

The CVT, YAMAGUCHI and TASCH models are designed as stand-alone models which incorporate all the effects required for simulating the carrier mobility.

As carriers are accelerated in an electric field their velocity will begin to saturate at high enough electric fields. This effect has to be accounted for by a reduction of the effective mobility since the magnitude of the drift velocity is the product of the mobility and the electric field component in the direction of the current flow. The following Caughey and Thomas expression [146] is usually used to implement a field-dependent mobility that provides a smooth transition between low-field and high-field behavior:

$$\mu_n(E) = \mu_{n0} \left[1 + \left(\frac{\mu_{n0}E}{v_{sat}^n} \right)^{\beta_n} \right]^{-1/\beta_n}, \qquad (B.15)$$

$$\mu_p(E) = \mu_{p0} \left[1 + \left(\frac{\mu_{p0}E}{v_{sat}^p} \right)^{\beta_p} \right]^{-1/\beta_p}, \qquad (B.16)$$

where E is the parallel electric field and μ_{n0} and μ_{p0} are the low-field electron and hole mobilities, respectively. The low-field mobilities are either set explicitly on the MOBILITY statement or calculated by one of the low-field mobility models. The model parameters $\beta_n = 2$ (BETAN) and $\beta_p = 1$ (BETAP) are user definable on the MOBILITY statement. The saturation velocities are calculated by default from the temperature dependent model [147]:

$$v_{sat}^n = v_{sat}^p = \frac{2.4 \times 10^7}{1 + 0.8 \exp\left(\frac{T_L}{600} \right)} [\text{cm s}^{-1}] \qquad (B.17)$$

but can be set to constant values on the MOBILITY statement in Silvaco ATLAS using the parameters VSATN and VSATP. In this case no temperature dependence is implemented. Specifying the FLDMOB parameter on the MODELS statement of the Silvaco ATLAS simulation software invokes the field-dependent mobility. FLDMOB should always be specified unless one of the inversion layer mobility models (which incorporate their own dependence on the parallel field) are specified.

References

[1] D. K. Ferry and S. M. Goodnick, *Transport in Nanostructures* (Cambridge Studies in Semiconductor Physics and Microelectronic Engineering, 1997).

[2] D. Vasileska and S. M. Goodnick, *Materials Science and Engineering, Reports: A Review Journal*, R38, no. 5, pp. 181–236 (2002).

[3] S. M. Goodnick and D. Vasileska, "Computational electronics," in *Encyclopedia of Materials: Science and Technology*, vol. 2, K. H. J. Buschow, R. W. Cahn, M. C. Flemings, E. J. Kramer and S. Mahajan Eds. (Elsevier, New York, 2001), pp. 1456–1471.

[4] A. Schütz, S. Selberherr, and H. Pötzl, "A two-dimensional model of the avalanche effect in MOS transistors," *Solid State Electron.*, vol. 25, pp. 177–183 (1982).

[5] P. Antognetti and G. Massobrio, *Semiconductor Device Modeling with SPICE* (McGraw-Hill, New York, 1988).

[6] M. Shur, *Physics of Semiconductor Devices* (Prentice Hall Series in Solid State Physical Electronics, New Jersey, 1990).

[7] D. L. Scharfetter and D. L. Gummel, "Large signal analysis of a Silicon Read diode oscillator," *IEEE Trans. Electron. Devices*, vol. ED-16, pp. 64–77 (1969).

[8] K. Bløtekjær, "Transport equations for electrons in two-valley semiconductors," *IEEE Trans. Electron. Dev.*, vol. 17, p. 38 (1970).

[9] M. V. Fischetti and S. E. Laux, "Monte Carlo simulation of submicron Si MOSFETs," in *Simulation of Semiconductor Devices and Processes*, vol. 3, G. Baccarani and M. Rudan Eds. (Technoprint, Bologna, 1988), pp. 349–368.

[10] L. V. Keldysh, *Sov. Phys.—JETP*, vol. 20, p. 1018 (1965).

[11] A. L. Fetter and J. D. Walecka, *Quantum Theory of Many-Particle Systems* (McGraw-Hill, New York, 1971).

[12] G. D. Mahan, *Many-Particle Physics* (Kluwer Academic/Plenum Publishers, New York, 2000).

[13] R. Lake, G. Klimeck, R. C. Bowen, and D. Jovanovic, "Single and multiband modeling of quantum electron transport through layered semiconductor devices," *J. Appl. Phys.*, vol. 81, p. 7845 (1997).doi:10.1063/1.365394

[14] G. Baccarani and M. Wordeman, "An investigation of steady-state velocity overshoot in silicon," *Solid State Electron.*, vol. 28, p. 407 (1985).doi:10.1016/0038-1101(85)90100-5

[15] S. Cordier, *Math. Mod. Meth. Appl. Sci.*, vol. 4, p. 625 (1994). doi:10.1142/S0218202594000352

[16] C. Kittel, *Introduction to Solid State Physics* (Wiley, New York, 1986, 6th ed.).

[17] N. W. Aschroft and N. D. Mermin, *Solid State Physics* (Saunders College Publishing, Philadelphia, Pennsylvania, 1976).

[18] P. Y. Yu and M. Cardona, *Fundamentals of Semiconductors* (Springer-Verlag, Berlin, 1999).

[19] R. M. Martin, *Electronic Structure: Basic Theory and Practical Methods*, (Cambridge University Press, Cambridge).

[20] C. Herring, *Phys. Rev.*, vol. 57, p. 1169 (1940).doi:10.1103/PhysRev.57.1169

[21] D. J. Chadi and M. L. Cohen, *Phys. Stat. Sol. (b)*, vol. 68, p. 405 (1975).

[22] J. Luttinger and W. Kohn, *Phys. Rev.*, vol. 97, p. 869 (1955). doi:10.1103/PhysRev.97.869

[23] M. L. Cohen and T. K. Bergstresser, *Phys. Rev.*, vol. 141, p. 789 (1966). doi:10.1103/PhysRev.141.789

[24] J. R. Chelikowsky and M. L. Cohen, *Phys Rev. B*, vol. 14, p. 556 (1976). doi:10.1103/PhysRevB.14.556

[25] Due to the fact that the heavy hole band does not have a spherical symmetry there is a discrepancy between the actual effective mass for density of states and conductivity calculations (number on the right) and the calculated value (number on the left) which is based on spherical constant-energy surfaces. The actual constant-energy surfaces in the heavy hole band are "warped," resembling a cube with rounded corners and dented-in faces.

[26] C. Jacoboni and L. Reggiani, "The Monte Carlo method for the solution of charge transport in semiconductors with applications to covalent materials," *Rev. Mod. Phys.*, vol. 55, pp. 645–705 (1983).doi:10.1103/RevModPhys.55.645

[27] P. J. Price, "Monte Carlo calculation of electron transport in solids," *Semiconduct. Semimet.*, vol. 14, pp. 249–334 (1979).

[28] C. Jacoboni and P. Lugli, *The Monte Carlo Method for Semiconductor Device Simulation* (Springer-Verlag, New York, 1989).

[29] D. J. Griffits, *Introduction to Quantum Mechanics* (Prentice Hall Inc., Englewood Cliffs, New Jersey, 1995).

[30] D. K. Ferry, *Quantum Mechanics: An Introduction for Device Physicists and Electrical Engineers* (Institute of Physics Publishing, London, 2001).

[31] R. F. Pierret, *Semiconductor Device Fundamentals* (Addison-Wesley, Reading, Massachusetts, 1996).

[32] S. M. Sze, *Physics of Semiconductor Devices* (Wiley, New York, 1981).

[33] M. Lundstrom, *Fundamentals of Carrier Transport* (Cambridge University Press, Cambridge, 2000).

[34] B. K. Ridley, *Quantum Processes in Semiconductors* (Oxford University Press, Oxford, 1988).

[35] D. K. Ferry, *Semiconductor Transport* (Taylor & Francis, London, 2000).

[36] J. P. McKelvey, *Solid State and Semiconductor Physics* (Krieger, Malabar, Florida, 1982).

[37] J. M. Ziman, *Electrons and Phonons: The Theory of Transport Phenomena in Solids* (Oxford University Press, New York, 2001).

[38] L. I. Schiff, *Quantum Mechanics* (McGraw-Hill, New York, 1955).

[39] Y.-C. Chang, D. Z.-Y. Ting, J. Y. Tang, and K. Hess, *Appl. Phys. Lett.*, vol. 42, p. 76 (1983).doi:10.1063/1.93732

[40] L. Reggiani, P. Lugli, and A. P. Jauho, *Phys. Rev. B*, vol. 36, p. 6602 (1987). doi:10.1103/PhysRevB.36.6602

[41] D. K. Ferry, A. M. Kriman, H. Hida, and S. Yamaguchi, *Phys. Rev. Lett.*, vol. 67, p. 633 (1991).

[42] P. Bordone, D. Vasileska, and D. K. Ferry, *Phys. Rev. B*, vol. 53, p. 3846 (1996). doi:10.1103/PhysRevB.53.3846

[43] P. A. Markowich and C. Ringhofer, *Semiconductor Equations* (Springer-Verlag Wien, New York, 1990).

[44] C. Ringhofer, "Numerical methods for the semiconductor boltzmann equation based on spherical harmonics expansions and entropy discretizations," *Transp. Theory and Stat. Phys.*, vol. 31, pp. 431–452 (2002).doi:10.1081/TT-120015508

[45] www.NanoHub.org

[46] S. Selberherr, *Simulation of Semiconductor Devices and Processes* (Springer-Verlag Wien, New York).

[47] K. Tomizawa, *Numerical Simulation of Submicron Semiconductor Devices* (The Artech House Materials Science Library, 1993).

[48] H. K. Gummel, "A self-consistent iterative scheme for one-dimensional steady state transistor calculation," *IEEE Trans. Electron. Devices*, vol. 11, pp. 455–465 (1964).

[49] T. M. Apostol, *Calculus*, Vol. II, *Multi-Variable Calculus and Linear Algebra* (Blaisdell, Waltham, Massachusetts, 1969) Ch. 1.

[50] J. W. Slotboom, "Computer-aided two-dimensional analysis of bipolar transistors," *IEEE Trans. Electron. Devices*, vol. 20, pp. 669–679 (1973).

[51] A. DeMari, "An accurate numerical steady state one-dimensional solution of the p-n junction," *Solid State Electron.*, vol. 11, pp. 33–59 (1968).doi:10.1016/0038-1101(68)90137-8

[52] D. L. Scharfetter and D. L. Gummel, "Large signal analysis of a Silicon Read diode oscillator," *IEEE Transaction on Electron Devices*, vol. ED-16, pp. 64–77 (1969).

[53] K. K. Thornber, "Current equations for velocity overshoot," *IEEE Electron. Device Lett.*, vol. 3, pp. 69–71 (1982).

[54] E. C. Kan, U. Ravaioli, and D. Chen, "Multidimensional augmented current equations with velocity overshoot," *IEEE Electron. Device Lett.*, vol. 12, p. 419 (1991). doi:10.1109/55.119151

[55] E. C. Kan, D. Chen, U. Ravaioli, and R. W. Dutton, "Numerical characterization of a new energy transport model," in *Proceedings of the International Workshop on Computational Electronics*, Beckman Institute, Urbana, IL, May 28–29, 1992.

[56] J. G. Ruch, "Electron dynamics in short-channel field-effect transistors," *IEEE Trans. Electron. Devices*, vol. 19, p. 652 (1972).

[57] S. Chou, D. Antoniadis, and H. Smith, "Observation of electron velocity overshoot in sub-100-nm-channel MOSFETs in Si," *IEEE Electron. Device Lett.*, vol. 6, p. 665 (1985).

[58] G. Shahidi, D. Antoniadis, and H. Smith, "Electron velocity overshoot at room and liquid nitrogen temperatures in silicon inversion layers," *IEEE Electron. Device Lett.*, vol. 8, p. 94 (1988).

[59] S. Selberherr, *Analysis and Simulation of Semiconductor Devices* (Springer-Verlag, New York, 1984).

[60] G. K. Wachutka, "Rigorous thermodynamic treatment of heat generation and conduction in semiconductor device modeling," *IEEE Trans. Comput.-Aided Des.*, vol. 9, pp. 1141–1149 (1990).

[61] K. Hess, *Theory of Semiconductor Devices* (IEEE, Piscataway, New Jersey, 2000).

[62] J. D. Bude, "Impact ionization and distribution functions in sub-micron nMOSFET technologies," *IEEE Electron. Device Lett.*, vol. 16, pp. 439–441 (1995). doi:10.1109/55.464810

[63] B. Meinerzhagen and W. L. Engl, "The influence of the thermal equilibrium approximation on the accuracy of classical two-dimensional numerical modeling of silicon submicrometer MOS transistors," *IEEE Trans. Electron. Devices*, vol. 35, pp. 689–697 (1988).doi:10.1109/16.2514

[64] J. D. Bude, "MOSFET modeling into the ballistic regime," in *Proceedings of the Simulation Semiconductor Processes and Devices, 2000*, pp. 23–26.

[65] R. Straton, "Diffusion of hot and cold electrons in semiconductor barriers", *Phys. Rev.*, vol. 126, pp. 2002–2014 (1962).doi:10.1103/PhysRev.126.2002

[66] T. Grasser, T.-W. Tang, H. Kosina, and S. Selberherr, "A review of hydrodynamic and energy-transport models for semiconductor device simulation," *Proc. IEEE*, vol. 91, pp. 251–274 (2003).

[67] M. A. Stettler, M. A. Alam, and M. S. Lundstrom, "A critical assessment of hydrodynamic transport model using the scattering matrix approach," in *Proceedings of the NUPAD Conference*, pp. 97–102 (1992).

[68] A. Brandt, "Multi-level adaptive solutions to boundary value problems," *Math. Comput.*, vol. 31, pp. 333–390 (1977).doi:10.2307/2006422

[69] W. Hackbusch, *Multi-Grid Methods and Applications* (Springer-Verlag, New York, 1985).

[70] K. Stuben and U. Trottenberg, in *Multigrid Methods*, W. Hackbusch and U. Trottenberg, Eds., Springer Lecture Notes in Mathematics No. 960 (Springer-Verlag, New York, 1982), pp. 1–176.

[71] A. Brandt, in *Multigrid Methods*, W. Hackbusch and U. Trottenberg, Eds., Springer Lecture Notes in Mathematics No. 960 (Springer-Verlag, New York, 1982).

[72] L. Baker and C. More, *Tools for Scientists and Engineers* (McGraw-Hill, New York, 1991).

[73] W. L. Briggs, *A Multigrid Tutorial* (Philadelphia, S. I. A. M., 1987).

[74] D. Jesperson, *Multigrid Method for Partial Differential Equations* (Mathematical Association of America, Washington, 1984).

[75] S. F. McCormick, Ed., *Multigrid Methods: Theory, Applications and Supercomputing* (Marcel Dekker, New York, 1988).

[76] P. Wesseling, *An Introduction to Multigrid Methods* (Wiley, New York, 1992).

[77] D. P. Kennedy and R. R. O'Brien, "Computer Aided two-dimensional analysis of a planar type junction field-effect transistor," *IBM J. Res. Dev.*, vol. 4 (1970).

[78] J. W. Slotboom, "Iterative scheme for 1 and 2-dimensional d.c. transistor simulation," *Electron. Lett.*, vol. 5, pp. 677–678 (1969).

[79] N. Sano and A. Yoshii, "Yoshii 3D device simulations," *Phy. Rev. B*, vol. 45, p. 4171 (1992).

[80] S. E. Laux, M. V. Fischetti, and D. J. Frank, "Monte Carlo analysis of semiconductor devices: The DAMOCLES program," *IBM J. Res. Dev.*, vol. 34, pp. 466–494 (1990).

[81] Silvaco International, Santa Clara, CA, *ATLAS User's Manual*, Ed. 6, 1998.

[82] C. Jacoboni and L. Reggiani, *Rev. Mod. Phys.*, vol. 55, p. 645 (1983). doi:10.1103/RevModPhys.55.645

[83] C. Jacoboni and P. Lugli, *The Monte Carlo Method for Semiconductor Device Simulation* (Springer-Verlag, Vienna, 1989).

[84] K. Hess, *Monte Carlo Device Simulation: Full Band and Beyond* (Kluwer Academic Publishing, Boston, 1991).

[85] M. H. Kalos and P. A. Whitlock, *Monte Carlo Methods* (Wiley, New York, 1986).

[86] D. K. Ferry, *Semiconductors* (Macmillan, New York, 1991).

[87] H. D. Rees, *J. Phys. Chem. Solids*, vol. 30, p. 643 (1969).doi:10.1016/0022-3697(69)90018-3

[88] R. M. Yorston, *J. Comp. Phys.*, vol. 64, p. 177 (1986).doi:10.1016/0021-9991(86)90024-0

[89] S. Bosi S and C. Jacoboni, *J. Phys.*, vol. C 9, p. 315 (1976).

[90] P. Lugli and D. K. Ferry, *IEEE Trans. Elec. Dev.*, vol. 32, p. 2431 (1985).

[91] N. Takenaka, M. Inoue and Y. Inuishi, *J. Phys. Soc. Jap.*, vol. 47, p. 861 (1979). doi:10.1143/JPSJ.47.861

[92] S. M. Goodnick and P. Lugli, *Phys. Rev. B*, vol. 37, p. 2578 (1988). doi:10.1103/PhysRevB.37.2578

[93] M. Moško, A. Mošková, and V. Cambel, *Phys. Rev. B*, vol. 51, p. 16860 (1995). doi:10.1103/PhysRevB.51.16860

[94] L. Rota, F. Rossi, S. M. Goodnick, P. Lugli, E. Molinari, and W. Porod, *Phys. Rev. B*, vol. 47, p. 1632 (1993).doi:10.1103/PhysRevB.47.1632

[95] R. Brunetti, C. Jacoboni, A. Matulionis, and V. Dienys, *Physica B&C*, vol. 134, p. 369 (1985).

[96] P. Lugli and D. K. Ferry, *Phys. Rev. Lett.*, vol. 56, p. 1295 (1986). doi:10.1103/PhysRevLett.56.1295

[97] J. F. Young and P. J. Kelly, *Phys. Rev. B*, vol. 47, p. 6316 (1993). doi:10.1103/PhysRevB.47.6316

[98] R. W. Hockney and J. W. Eastwood, *Computer Simulation Using Particles* (Institute of Physics Publishing, Bristol, 1988).

[99] D. J. Adams and G. S. Dubey, *J. Comp. Phys.*, vol. 72, p. 156 (1987).doi:10.1016/0021-9991(87)90076-3

[100] Z. H. Levine and S. G. Louie, *Phys. Rev. B*, vol. 25, p. 6310 (1982). doi:10.1103/PhysRevB.25.6310

[101] L. V. Keldysh, *Zh. Eksp. Teor. Fiz.*, vol. 37, p. 713 (1959).

[102] N. Sano and A. Yoshii, *Phys. Rev. B*, vol. 45, p. 4171 (1992).

[103] M. Stobbe, R. Redmer and W. Schattke, *Phys. Rev. B*, vol. 47, p. 4494 (1994).

[104] Y. Wang and K. Brennan, *J. Appl. Phys.*, vol. 71, p. 2736 (1992).

[105] M. Reigrotzki, R. Redmer, N. Fitzer, S. M. Goodnick, M. Dür, and W. Schattke, *J. Appl. Phys.*, vol. 86, p. 4458 (1999).

[106] D. J. Chadi and M. L. Cohen, *Phys. Stat. Sol. (b)*, vol. 68, p. 405 (1975).

[107] M. V. Fischetti and S. E. Laux, *Phys. Rev. B*, vol. 38, p. 9721 (1988).

[108] For a complete overview, see www.research.ibm.com/DAMOCLES.

[109] M. Saraniti and S. M. Goodnick, *IEEE Trans. Elec. Dev.*, vol. 47, p. 1909 (2000).

[110] T. Gonzalez and D. Pardo, *Solid State Electron.*, vol. 39, p. 555 (1996).

[111] P. A. Blakey, S. S. Cherensky, and P. Sumer, *Physics of Submicron Structures* (Plenum Press, New York, 1984).

[112] T. Gonzalez and D. Pardo, *Solid-State Electron.*, vol. 39, p. 555 (1996).

[113] R. W. Hockney and J. W. Eastwood, *Computer Simulation Using Particles* (Institute of Physics Publishing, Bristol, 1988).

[114] S. E. Laux, *IEEE Trans. Comp.-Aided Des. Int. Circ. Syst.*, vol. 15, p. 1266 (1996).

[115] S. S. Pennathur and S. M. Goodnick, *Inst. Phys. Conf. Ser.*, vol. 141, p. 793 (1995).

[116] S. Beysserie, J. Branlard, S. Aboud, S. Goodnick, and M. Saraniti, *Superlatt. Microstruc.*, accepted for publication (2006).

[117] M. E. Kim, A. Das, and S. D. Senturia, *Phys. Rev. B*, vol. 18, p. 6890 (1978).

[118] M. V. Fischetti and S. E. Laux, *Phys. Rev. B*, vol. 38, p. 9721 (1988).

[119] P. Lugli and D. K. Ferry, *Phys. Rev. Lett.*, vol. 56, p. 1295 (1986).

[120] A. M. Kriman, M. J. Kann, D. K. Ferry, and R. Joshi, *Phys. Rev. Lett.*, vol. 65, p. 1619 (1990).

[121] R. P. Joshi and D. K. Ferry, *Phys. Rev. B*, vol. 43, p. 9734 (1991).

[122] M. V. Fischetti and S. E. Laux, *J. Appl. Phys.*, vol. 78, p. 1058 (1995).

[123] W. J. Gross, D. Vasileska, and D. K. Ferry, *IEEE Electron. Device Lett.*, vol. 20, p. 463 (1999).

[124] W. J. Gross, D. Vasileska, and D. K. Ferry, VLSI Design 10, 437 (2000).

[125] D. Vasileska, W. J. Gross, and D. K. Ferry, *Superlatt. Microstruc.*, vol. 27, p. 147 (2000).

[126] W. J. Gross, D. Vasileska, and D. K. Ferry, *IEEE Trans. Electron Dev.*, vol. 47, p. 1831 (2000).

[127] K. Tomizawa, *Numerical Simulation of Submicron Semiconductor Devices* (Artech House, Inc., Norwood, 1993).

[128] H. Brooks, *Phys. Rev.*, vol. 83, p. 879 (1951).

[129] C. Canali, G. Ottaviani and A. Alberigi-Quaranta, *J. Phys. Chem. Solids*, vol. 32, p. 1707 (1971).

[130] G. Strang, *Linear Algebra and Its Applications* (Academic Press, New York, 1980, 2nd ed.).

[131] G. V. Gadiyak and M. S. Obrecht, *Simulation of Semiconductor Devices and Processes: Proceedings of the Second International Conference*, 147 (1986).

[132] H. L. Stone, *SIAM J. Numer. Anal.*, vol. 5, p. 536 (1968).

[133] W. Hackbush, *Multi-Grid Methods and Applications* (Springer-Verlag, Berlin, 1985).

[134] A. Brandt, "Multi-level adaptive technique (MLAT): The Multi-grid Method", *IBM Research Report RD-6026*, IBM T. J. Watson Research Center, Yorktown Heights, New York, 1976.

[135] R. E. Bank and D. J. Rose, "An O(N*N) method for solving constant coefficient boundary value problems in two dimensions," *SIAM J. Numer. Anal.*, vol. 12, pp. 529–540 (1975).

[136] D. K. Schroder, *Semiconductor Material and Device Characterization* (John Wiley & Sons, Inc., New York, 1990).

[137] D. K. Schroder, *Modular Series on Solid State Devices: Advanced MOS Devices* (Addison-Wesley Publishing Company, New York, 1987).

[138] D. M. Caughey R. E. Thomas, "Carrier mobilities in silicon empirically related to doping and field", *Proc. IEEE*, vol. 55, pp. 2192–2193 (1967).

[139] N. D. Arora, J. R. Hauser, and D. J. Roulston, "Electron and hole mobilities in silicon as a function of concentration and temperature," *IEEE Trans. Electron. Devices*, vol. 29, pp. 292–295 (1982).

[140] J. M. Dorkel and PH. Leturcq, "Carrier mobilities in silicon semi-empirically related to temperature, doping and injection level", *Solid State Electron.*, vol. 24, pp. 821–825 (1981).doi:10.1016/0038-1101(81)90097-6

[141] D. B. M. Klaassen, "A unified mobility model for device simulation—I. Model equations and concentration dependence," *Solid State Electron.*, vol. 35, pp. 953–959 (1992). doi:10.1016/0038-1101(92)90325-7

[142] M. Shirahata, H. Kusano, N. Kotani, S. Kusanoki, and Y. Akasaka, "A mobility model including the screening effect in MOS inversion layer", *IEEE Trans. Comp.-Aided Design*, vol. 11, pp. 1114–1119 (1988).

[143] C. Lombardi, S. Manzini, A. Saporito, and M. Vanzi, "A physically based mobility model for numerical simulation of nonplanar devices," *IEEE Trans. Comp.-Aided Design*, vol. 7, pp. 1154–1171 (1992).

[144] Ken Yamaguchi, "Field-dependent mobility model for two-dimensional numerical analysis of MOSFET's," IEEE Trans. Electron Devices, vol. 26, pp. 1068–1074 (1979).

[145] G. M. Yeric, A. F. Tasch, and S. K. Banerjee, "A universal MOSFET mobility degradation model for circuit simulation", *IEEE Trans. Computer-Aided Design*, vol. 9, p. 1123 (1991).doi:10.1109/43.62736

[146] D. M. Caughey and R. E. Thomas, "Carrier mobilities in silicon empirically related to doping and field", *Proc. IEEE*, vol. 55, pp. 2192–2193 (1967).

[147] *Silvaco ATLAS User Manual*. Available at: www.silvaco.com.

Author Biography

Dragica Vasileska received the B.S.E.E. (Diploma) and the M.S.E.E. Degree form the University Sts. Cyril and Methodius (Skopje, Republic of Macedonia) in 1985 and 1992, respectively, and a Ph.D. Degree from Arizona State University in 1995. From 1995 until 1997 she held a Faculty Research Associate position within the Center of Solid State Electronics Research at Arizona State University. In the fall of 1997 she joined the faculty of Electrical Engineering at Arizona State University. In 2002 she was promoted to Associate Professor with tenure. Her research interests include semiconductor device physics and semiconductor device modeling, with strong emphasis on quantum transport and Monte Carlo particle-based simulations. She is a member of IEEE and APS. Dr. Vasileska has published more than 100 journal publications, over 80 conference proceedings refereed papers, has given numerous invited talks and is a co-author on a book on Computational Electronics with Prof. S. M. Goodnick. She has many awards including the best student award from the School of Electrical Engineering in Skopje since its existence (1985, 1990). She is also a recipient of the 1998 NSF CAREER Award. Her students Ashwin Ashok and Santhosh Krishnan have won the best presentation and the best poster award at the LDSD conference in Cancun, 2004.

Steve Goodnick is Associate Vice President for Research at Arizona State University. His research specializations lie in solid-state device physics, semi-conductor transport, quantum and nanostructure devices and device technology, and high frequency devices. He will maintain his leadership of ASU's nanoelectronics efforts as director while in this post.

Goodnick previously served as the interim deputy dean for the Ira A. Fulton School of Engineering at ASU, and earlier as chair of the Fulton School's Department of Electrical Engineering, one of ASU's most active and successful units, and served as President of the Electrical and Computer Engineering Department Heads Association from 2003–2004. He received his B.S. in engineering science from Trinity University in 1977, and his M.S. and Ph.D. degrees in electrical engineering from Colorado State University in 1979 and 1983, respectively. Germany, Japan and Italy are among the countries he has served as a visiting scientist.

Goodnick is a Fellow of the Institute of Electrical and Electronics Engineers (IEEE) and an Alexander von Humboldt Research Fellow. Other honors and awards he has received include the IEEE Phoenix Section Society Award for Outstanding Service (2002), the Colorado State University College of Engineering Achievement in Academia Award (1998), and the College of Engineering Research Award (Oregon State University, 1996). He is a member of IEEE, the American Physical Society, the American Association for the Advancement of Science, and the American Society of Engineering Education. His publication record includes more than 165 refereed journal articles, books and book chapters related to transport in semiconductor devices and microstructures.